W0081977

Rethinking the Three R's in Animal Research

Jan Lauwereyns

Rethinking the Three R's in Animal Research

Replacement, Reduction, Refinement

palgrave
macmillan

Jan Lauwereyns
Kyushu University
Fukuoka, Japan

ISBN 978-3-319-89299-3   ISBN 978-3-319-89300-6   (eBook)
https://doi.org/10.1007/978-3-319-89300-6

Library of Congress Control Number: 2018938322

© The Editor(s) (if applicable) and The Author(s) 2018
This work is subject to copyright. All rights are solely and exclusively licensed by the
Publisher, whether the whole or part of the material is concerned, specifically the rights
of translation, reprinting, reuse of illustrations, recitation, broadcasting, reproduction
on microfilms or in any other physical way, and transmission or information storage and
retrieval, electronic adaptation, computer software, or by similar or dissimilar methodology
now known or hereafter developed.
The use of general descriptive names, registered names, trademarks, service marks, etc. in this
publication does not imply, even in the absence of a specific statement, that such names are
exempt from the relevant protective laws and regulations and therefore free for general use.
The publisher, the authors and the editors are safe to assume that the advice and
information in this book are believed to be true and accurate at the date of publication.
Neither the publisher nor the authors or the editors give a warranty, express or implied,
with respect to the material contained herein or for any errors or omissions that may have
been made. The publisher remains neutral with regard to jurisdictional claims in published
maps and institutional affiliations.

Cover illustration: © Bitboxx.com

Printed on acid-free paper

This Palgrave Pivot imprint is published by the registered company Springer International
Publishing AG part of Springer Nature
The registered company address is: Gewerbestrasse 11, 6330 Cham, Switzerland

To MAPW

CONTENTS

CHAPTER 1

Introduction

Abstract Anyone involved in animal research knows the three R's—Replacement, Reduction, and Refinement. Russell and Burch introduced these groundbreaking principles more than sixty years ago; now, it is time for a reassessment with the aim of improving the ways in which we conduct animal research. The opening chapter lays out the basic premise and perspectives of the book, with brief previews of the chapters.

Keywords Animal research · Animal ethics · Guiding principles

I think we can do better in many areas of life. Here, in this short book, I concentrate on the room for improvement in animal research. I am writing this small book of critique for you, dear animal investor (that is what I will call you: *animal investor*). An animal investor may be any person (or human being, *Homo sapiens*) who is actively concerned with animal research—whether it be as a practitioner, a policymaker, or an advocate (for or against). This book provides new ideas and proposals based on the so-called three R's—Replacement, Reduction, and Refinement—which were first formulated by W.M.S. ("Bill") Russell and R.L. ("Rex") Burch in their classic text, *The Principles of Humane Experimental Technique* (1959/1992). My objective in this book is simple: I aim to explain how we can improve the ethics of animal research. Thus, I offer a scholarly work of thought before action. (The action would have to be joint action, and will happen only if you agree.)

© The Author(s) 2018
J. Lauwereyns, *Rethinking the Three R's in Animal Research*,
https://doi.org/10.1007/978-3-319-89300-6_1

1

1.1   Building on the Legacy of Russell
and Burch (1959/1992)

You do not have to have read Russell and Burch to be able to join the discussion. Most animal investors have not read Russell and Burch. In fact, the book is out of print (1992 reprints are available for roughly 300 US dollars—a price I found too steep, even though I can promise you I am *really* interested in the topic). An online version exists (courtesy of ALTWEB, "the global clearinghouse for information on alternatives to animal testing"), which should technically be regarded as "accessible"; however, practically, it is difficult to read and requires many clicks. I eventually managed to read many sections of it (diligently, forcing myself, while preparing for this book), but I strongly believe animal investors can benefit from the book without having to read it. Russell and Burch's classic has served its purpose: it created a legacy. The three R's have entered the mainstream in the practice and policymaking regarding animal research. Yet, the mainstream did not absorb the original definitions. Instead, if we take a closer look at how the three R's are being interpreted and applied, we quickly find that the concepts have morphed significantly in their real-world afterlife.

Most animal investors have not read Russell and Burch (1959/1992), but very many of us can name the three R's. In doing so, we understand them intuitively, in our own ways, often without actually defining them. Here, I aim to rethink the three R's by pointing out some issues with the original definitions, as well as with some of the later interpretations and applications. The book is structured with four movements to develop a critical assessment of the validity of the three R's as guiding principles for animal research. To go straight to the core of the argument, I pull Replacement into the foreground as the one principle that requires full priority; in fact, it may be the only purely ethical principle among the three R's. (When emphasizing the priority of Replacement, I use a revised definition that is broader than the original.) Reduction and Refinement, on the other hand, may have more to do with balancing the economics and quality of science than with ethics. I argue that, if Replacement fails, the proper ethical course of action is to rely on collective decision-making. Only through agreement can we make an ethically valid (utilitarian) case for the need of animal research.

1.2   Four Movements to the Argument

The first movement, which takes place in Chapter 2, considers the historical background and achievements of the work by Russell and Burch (1959/1992). I carefully examine the original definitions of the three R's, noting how the principles have come to occupy a central position that is broadly accepted by all interested parties. I describe the aims and rationale offered by Russell and Burch, situating these in the context of the late 1950s. I argue that the principles are products of their time, completely in line with the then-prevailing knowledge about and attitudes toward animals in society. A particular point of interest is the concept of "humanity" as applied in the formulation of the three R's. This is contrasted with more recent ideas about "speciesism" and the moral status of nonhuman animals. Today, the majority of people believe that animals deserve some amount of protection. This seems to imply a cost that weighs in opposition to potential benefits, such as those promised by science. As a case study, I focus on the use of nonhuman primates in basic neuroscience research, from the 1960s to the present. This historical analysis illustrates how the changing context implies a changing perspective on the ethical validity of research. Yet, the three R's fail as guidelines to translate the changes in context to commensurate changes in conduct. We must analyze this dysfunction of the three R's.

Chapter 3 presents the second movement in the argument, focusing on a micro-macro analysis of animal research. Viewed at the microscopic level of each individual laboratory, it may appear that researchers work in accordance with the three R's, choosing the appropriate animal model, aiming to get the smallest sample size needed for any experiment, and continuing to improve the techniques and knowledge extraction. However, turning to the macroscopic level of the entire field, we find mismatches between the individual intentions and overall outcomes. This chapter offers an analysis inspired by the groundbreaking work of Thomas C. Schelling and others on emergent patterns of macro-behavior as a function of micro-motives. Here, I anticipate that the key challenge will be to shift from agency at the microscopic level to agency at the macroscopic level. Such a shift would enable researchers to address, among other things, the key scientific problem of sample size. Following the micro-macro analysis, I offer my revised definitions of the three R's as principles toward reasonable experimental inquiry, to be orchestrated by the research community.

Chapter 4, home to the third movement in the argument, presents the use of monkeys in research as a particular area of controversy, with a critique of typical arguments offered in public debate. I note the weaknesses of retrospective thinking, the erroneous appeal to necessity, and the unfounded bias against working with human volunteers and rodents. The issues are examined more specifically via three scientific paradigms to illustrate the impasse with the three R's as guiding principles. One paradigm regards work on the development of a neural control system for robotic arm movement. As a second paradigm, I discuss work on the topic of perceptual decision-making with rats as a putative replacement for research with nonhuman primates. A third paradigm focuses on the topic of the cognitive mechanisms underlying cocaine addiction and relapse in nonhuman primates. The three paradigms allow us to reflect on the resistance by individual researchers, using their limited perspective to put forth misguided claims of irreplaceability of their preferred animal model.

Drawing on the analyses of the previous chapters, I construct an integrative view in Chapter 5, with clear and realistic proposals for the policymaking with respect to the use of animals in research. The aim is to come to a straightforward, internally coherent approach to the use of animals in research, which represents an upgrade of the three R's. Replacement comes first, as an inherently ethical principle. When Replacement is not possible, the use of animals should be carefully managed on the basis of collective decision-making—that is, at the level of research communities, universities, and funding agencies, but not at the level of individual researchers. This proposal is strengthened through a discussion in terms of opportunity costs (the alternative opportunities that are given up when engaging in a given research project); these are precisely the type of costs that often remain out of consideration when individual researchers design their research projects as autonomous agents, vulnerable to conflicts of interest. I argue that collective decision-making is necessary to ensure that we invest our time, money, and effort in the most optimal way.

The proposal essentially comes down to a macroscopic approach to managing research—an effort in line with recent calls for open science and mega-science, analogous to the Large Hadron Collider and the Human Genome Project. Collective decision-making enables communities to engage in particular animal research projects that seem crucial for important research progress that cannot be obtained through alternative means. Detailed suggestions are made on how to organize the macroscopic approach, through extant research communities and institutes, thinking with the concepts of open science and big science.

1.3 FULL DISCLOSURE

At the outset of this scholarly work, I want to note a few particularities about the position from which I write (or about "my voice," to use a poetical concept). Writing about ethics, I find, requires more than the usual level of attention to the personal dimension, including who I am, where I am coming from, what I know from experience, what I feel, my potential biases, and my prejudices. Thus, relevant information to know about the present writer may include the following.

I am an atheist. I do not adhere to any systematic moral philosophy. Furthermore, I find myself continuously grappling with the concepts of welfare, freedom, and virtue (to name the three approaches to justice, as listed by Sandel, 2009)—thinking I would like all three, but I do not always know what they are or how they line up (or not).

I value science very highly and believe that a knowledge-based society is necessary for welfare, freedom, and virtue.

I think democracy is generally the appropriate form of society. It tends to promote welfare, freedom, and virtue. This is a somewhat shaky thought because democratic processes can occasionally produce adverse effects. An infamous example is Nazi Germany, where the people put Hitler in power. Thus, we must watch over democracy with our own critical devices, speaking up when we note problems; in this vein is my current writing.

I value many forms of life and being (and even inorganic objects) very highly. However, I also have weaknesses: I find myself consuming certain forms of animal protein, from cheese to beefsteak, albeit in moderate amounts.

I do not think that humans have any sacred right to dominate or exploit other creatures (whether fellow humans or other animals).

I think there is good scientific evidence to suggest that humans have a particularly rich sentient life (or consciousness), to varying degrees comparable to other types of sentient life (very comparable to that of chimpanzees, less so to that of roundworms).

I think many issues of ethics are tied to aspects of sentient life: desires and feelings—from pain to pleasure, joy to suffering.

For the first twenty-nine years of my life, I was not particularly concerned with the use of animals, in research or otherwise. In my youth, my family had a number of dogs. Sometimes, I walked Chi-Chi. I used to play with insects, in games they typically did not survive (occasionally an animal took revenge; I remember a particularly instructive sting I received from a wasp I had tried to capture in a plastic bottle when I was eight years old). As already implied, I loved—and still do love—red meat (although these days, I try to eat it less often and in smaller portions than before my "ethical awakening").

Trained as a cognitive psychologist who studied the processes of visual attention through behavioral experiments with humans, I became interested in neurophysiology. I wanted to study the brain mechanisms of visual attention. In 1998, I had the opportunity to start working as a post-doctoral researcher in a neurophysiology laboratory led by Okihide Hikosaka, at Juntendo University in Japan. This was my first experience conducting research with animals, with Japanese macaques. After suppressing an initial emotional shock at some of the procedures, I continued working with monkeys for five years, trying to adapt to the research culture.

Gradually, I became more skeptical of the underlying ethical reasoning. In the fall of 2002, I decided to stop working with nonhuman primates. Instead, I switched to working with rodents and began studying animal ethics and animal welfare. I was trying to understand and properly evaluate my misgivings about the work with monkeys, aiming to construct a coherent vision on how to improve the praxis of research with animals. Since leaving "the monkey field" but still engaged in animal research, I have served on various animal ethics committees in New Zealand and Japan and have participated in numerous (mostly informal) debates, discussions, and interviews. I found myself frequently at odds with both scientists and animal rights activists, as I naturally tend to what sometimes appears to be the great desert (or No-Go Zone) of the middle ground. Committed to the progress of science and realizing the indispensable role of animals therein, I am also (apparently unusually) critical of how the research is currently being designed and implemented. With this book, I am offering my thoughts on the matter, comprehensively and as clearly as I can formulate them.

My thoughts have not emerged in isolation. So many friends, colleagues, scientists, writers, thinkers, doers, people, and animals have taught me so many valuable things, it would be impossible for me to name all the relevant names. Inevitably, there will be unforgivable omissions in any list, so I should avoid listing and put out an abstract IOU. Still, I feel compelled to mention a few names of creatures that have had, and continue to have, a particular impact. In the very first instance, I think of Haruki, the monkey I worked with from the Summer of 1998 to the first, cold days of 2001, when I was required, but unable, to kill him. Haruki, I can safely say, changed my life. I also think of monkeys Zola, Jun, Ryu, Shin, Ichitaro, Hope, Gauss, and especially Armstrong, whose will was stronger than mine. Some researchers have been supportive of my freedom of opinion, despite contrary views. Thanks go especially to Masamichi Sakagami; his unwavering friendship through (sometimes very radical) disagreement continues to baffle and delight me. He is peerless in this regard. For my perspective on animal ethics, some voices of encouragement and inspiration (past and present) stand out in my mind; they are those of Mariella Debille, Peter De Graef, Ulrike Draesner, Fernanda Ferreira, Dave Harper, Keiji Iramina, Ernest Koh, Hideo Kubo, Yuji Soejima, Heidi Thomson, Michel Vandenbosch, and the late Leo Vroman. Closer to home (a virtual or elusive kind of home), I hear the voices of my sisters Kim and Nathalie, my daughter Nanami, my son Shinsei, and, most irrationally, my unnamable Miki.

I acknowledge, very gratefully, the institutional support from Kyushu University (Fukuoka, Japan), particularly the Graduate School of Systems Life Sciences, the Faculty of Arts and Science, and the School for Interdisciplinary Science and Innovation; here I found a suitable context to expand the scope of my academic work, both in teaching and in research, to the field of bioethics. I also received support from the University of Leuven (KU Leuven, Leuven, Belgium), my alma mater. My thanks go particularly to (former Rector) Rik Torfs and Johan De Tavernier for organizing a symposium and debate on nonhuman primates in biomedical research, with a very good sample of the diversity of opinion, on 3 December 2014. That occasion greatly stimulated my thinking on the topic, with the present book as an outcome.

I would like to thank Rachel Daniel, my editor at Palgrave Macmillan, who made the project move forward with infectious conviction from day one, and Kyra Saniewska, who provided the perfect editorial assistance.

Finally, I would like to pay special tribute here to Ms. R.M. Bossuyt (a lawyer who has pledged her estate to biomedical research at the University of Leuven, in Belgium) for encouraging me—almost demanding me—to raise my voice when sometimes it seemed in my best interest not to do so. We have the politest correspondence, in very formal Dutch. Ms. Bossuyt starts every e-mail message to me with "Professor." I start every e-mail message to Ms. Bossuyt with *"Meester"*—literally "Master," but it is translated more appropriately as "Your Honor." This book is for you, Your Honor.

REFERENCES

Russell, W. M. S., & Burch, R. L. (1959/1992). *The Principles of Humane Experimental Technique*. Wheathampstead: Universities Federation for Animal Welfare. Available at: ALTWEB: http://altweb.jhsph.edu/pubs/books/humane_exp/foreword.

Sandel, M. J. (2009). *Justice: What's the Right Thing to Do*. New York: Farrar, Straus and Giroux.

CHAPTER 2

Concept Fatigue with the Three R's

Abstract The three R's as defined by Russell and Burch have come to occupy a central position in animal research. A careful examination of the original definitions and aims shows these to be a product of their time, no longer in line with contemporary perspectives. A particular point of interest is the reliance on the concept of "humanity." This is contrasted with more recent ideas about "speciesism" and the moral status of non-human animals. As a case study, I focus on the use of nonhuman primates in basic neuroscience research; the historical analysis illustrates how the changing context implies a changing perspective on the ethical validity of research. Yet, the three R's fail as guidelines to translate the changes in context to commensurate changes in conduct.

Keywords Animal ethics · Replacement · Reduction · Refinement Historical analysis

The three R's of Replacement, Reduction, and Refinement are a proper point of departure for any discussion of the ethics of animal research. It is routinely noted that most animal investors subscribe to Russell and Burch's (1959/1992) principles. A few quotes may suffice to set the tone. The first sentence of an article by Buchanan-Smith et al. (2005, p. 379) stated: "The principles behind the Three R's ... for animal experimentation ... are now widely accepted, and are fundamental to the philosophy underlying the guidelines and legislation that regulate animal

© The Author(s) 2018
J. Lauwereyns, *Rethinking the Three R's in Animal Research*,
https://doi.org/10.1007/978-3-319-89300-6_2

experimentation." The first two lines of the abstract from a paper by Olsson et al. (2012, p. 333) observed: "Over the 50 years since they were first proposed, the 3 R's … have made a tremendous impact. These principles seem to unify concerns for better science with causing less harm to animals." Tannenbaum and Bennett (2015, p. 120) wrote: "For years, these terms have appeared in virtually every context relating to the use of animals in research—in laws, regulations, and government policies; ethical pronouncements of professional research organizations; and books and journal articles."

Even researchers who advocate for the use of nonhuman primates in basic neuroscience research are happy to concede that the three R's are the stuff of consensus: "Animal protection laws … are built on the broad consensus across science, politics, and society that a certain amount of research on animals is necessary and justifiable. This consensus includes the 3R principles … of Replace, Refine, and Reduce" (Roelfsema and Treue 2014, p. 1200). The broad consensus, however, does nothing to resolve the vigorous debates between the different camps, for and against animal research. Something must be lost in translation. The three R's may seem easy to agree to at first, but putting them into practice turns out to be a problematic exercise.

Different people hear different things in the three R's. Sometimes, it sounds as if there are too many R's; thus, one R may be used as an excuse to forget about the others. Balls (2010, p. 21) intuited, "There is … a danger that refinement can be used as a convenient way of showing commitment to the Three R's, whilst ensuring that experimentation is seen as respectable." More recently, in the *Science* issue of 27 October 2017, I found clear signs of concept fatigue with the three R's in a news report with the promising, but ultimately depressing, title of "Revamp Animal Research Rules, Report Urges" (Cornwall 2017, p. 434). Apparently, the Association of American Medical Colleges and like-minded groups are lobbying for less oversight and a smaller administrative burden in animal research. Sally Thompson-Iritani, a member of one of the groups, was quoted saying that, "Scientists chafe at rules requiring them to check the literature for less invasive alternatives" (ibid.). In a bid to cut red tape, the groups call for doing away with literature searches for alternatives to testing. In the margin, I put a big exclamation mark. The goal, according to Thompson-Iritani, "is getting scientists back to the bench doing their research, and animal care specialists getting back to their animals" (ibid.). This is a surprisingly frank (even somewhat refreshing, if shocking) admission that responding to the R of Replacement is seen as a waste of time—a useless search for an answer that is always set in stone: Replacement will not be possible. To my eyes, this looks like a textbook example of prejudice. Perhaps all is not well with the three R's. Perhaps we should re-examine the original definitions and learn how the broad consensus in favor of the three R's does not reflect a true agreement, but rather conflicting interpretations.

2.1   THE ORIGINAL WORK BY RUSSELL AND BURCH (1959/1992) IN CONTEXT

2.1.1   On the Removal of Inhumanity

In the preface of *The Principles of Humane Experimental Technique*, Russell and Burch (1959/1992) acknowledged the support of the Universities Federation of Animal Welfare. (Unfortunately, the online publication of the book does not have any pagination; for convenience, let me refer to it as *PHET*, listing in each case the title of the relevant section that is accessible as a separate page on ALTWEB; thus, here, *PHET, Preface*.) The Universities Federation of Animal Welfare is registered as a United Kingdom scientific and educational charity and is still in operation today (see https://www.ufaw.org.uk). Unclear is whether the work by Russell and Burch, a five-year long "systematic research on the progress of humane technique in the laboratory" (*PHET*, ibid.), was actually commissioned by the society or whether the pair had been awarded a grant upon submitting a proposal. In any case, the two researchers received the backing of a respectable society, with a decades-long history of a double commitment to science and animal welfare. In addition to financial support, Russell and Burch gained advice on their research from a "special Consultative Committee" (*PHET*, ibid.), chaired by Professor P.B. Medawar. The work did not come out of the blue, but represented a significant moment in a very active development of ideas and practices with respect to animal experimentation. It built on a tradition that in 1959 was already more than 80 years old, dating back to before the *Cruelty to Animals Act 1876*.

In the opening paragraph of their introductory chapter, Russell and Burch noted the "vast increase in the numbers of nonhuman animals employed as the subjects of experiment," numbers "reckoned annually in millions" and "rising steadily" (*PHET, Chapter 1, Scope of the Study*). Right after this observation, Russell and Burch claimed it to be a "truism" that "we owe to animal experimentation many if not most of the benefits of modern medicine and countless advances in fundamental scientific knowledge" (*PHET*, ibid.). To the notion that this would involve "an irreconcilable conflict" between science and the humane treatment of animals, Russell and Burch offered a firm rebuttal: "The conflict disappears altogether on closer inspection, and by now it is widely recognized that the humanest possible treatment of experimental animals, far from being an obstacle, is actually a prerequisite for successful animal experiments," and "the intimate relationship between humanity and efficiency in experimentation will recur constantly as a major theme in the present book" (*PHET*, ibid.).

In a footnote to this opening paragraph, Russell and Burch noted, "the word 'humanity' is used in its secondary sense of 'humaneness'" (*PHET, Chapter 1, Scope of the Study*). Yet the word *humanity* inevitably rings through, also in its primary sense. What aspects of *Homo sapiens* are being referred to in this "humanity" and "humaneness"? Presumably, our thoughts should not be directed to the vicious elements in human nature, which have prompted some authors to designate ours as a killer species (Wrangham 2004). Just to be sure, I checked the adjective *humane* in the online Oxford English Dictionary (https://www.oed.com): It means "civil, courteous, or obliging towards others," in later use characterized by "sympathy with and consideration for others; feeling or showing compassion towards humans or animals; benevolent, kind."

On the surface of it, Russell and Burch appeared to be concerned with a form of animal experimentation that would be compatible with "humanity"—the good-natured aspects of being human. Inhumanity, then—as the absence of humanity—mismatches with our (supposedly) intrinsic good nature. Incidentally, in the Dutch language, cruel experiments on animals are sometimes designated as *mensonterend* (which is usually translated as "humiliating," but literally specifies that the humiliation is suffered by humans). This, I think, is a particularly "speciesist" concern, to use the term (analogous to racism) coined by Richard D. Ryder and popularized by Peter Singer in *Animal Liberation* (1975/2009): a prejudice placing humans, principally and categorically, above animals. The way Russell and Burch phrased it, causing suffering to an animal implies an insult—not to the animal, but to humanity.

2.1.2   *Humane Versus Speciesist*

My reading of Russell and Burch's concern with humanity as something speciesist, I argue, matches perfectly with how they used their truism with respect to the medical and scientific progress thanks to animal experimentation (which, I will show, reflects another instance of speciesism). It is unquestionable that animal experimentation has led to progress, claimed Russell and Burch. However, they proceeded to leave the progress unquestioned. Coming from research scientists, such usage of the truism is disappointingly unscientific—a blanket statement that preempts any request for evidence-based utility of the research. Effectively, Russell and Burch did not bother examining the utility of any specific research project. Instead, they relied on the truism in an overgeneralizing way. Let me paraphrase it in the form of a syllogism:

Major premise: *Animal experimentation leads to progress.*
Minor premise: *This is a case of animal experimentation.*
Therefore: *This case of animal experimentation will lead to progress.*

I hope it will be obvious that this reasoning is erroneous. The truism, properly stated as a fact, is that *some* animal experiments *have produced* progress (in the past). Other animal experiments have not. Moreover, past success is not a guarantee for future success. Finally, past success in one way does not prove that the same success could not be obtained in another way. It is anything but a given that a new animal experiment will lead to a justifiable level of progress.

How could Russell and Burch have made such a glaring mistake in simple logic? Their reasoning was not obviously mistaken in the context of their time. It was speciesist at a time when the term had not yet been coined. Perhaps the two researchers were good Christians or followers of another Abrahamic or other religion that principally and categorically places humans above other animals. (Speciesists would take offense at my usage of the adjective *other* and its biologically valid implication that humans are a type of animal.) According to speciesists, any human benefit, however unlikely or small, would always and necessarily take precedence over any animal cost, however massive. Thinking like a speciesist, it makes perfect sense that—given the truism that animal experimentation leads to medical or scientific progress—any new animal experiment is justified.

Russell and Burch suggested that we should just concern ourselves with avoiding "inhumanity" in the technique. To be humane, here, conforms to the secondary meaning, as given by the Oxford English Dictionary—that is, behavior "designed or calculated to inflict minimal pain." Tannenbaum and Bennett (2015), in their excellent linguistic analysis, pointed out that in *PHET* the term *inhumanity* is used descriptively, not normatively, and simply stands for "distress." By avoiding inhumanity, the efficiency of our data collection would be improved. In Russell and Burch's view, it is simply a matter of good science. Right before actually defining the three R's of Replacement, Reduction, and Refinement, they included a section on *The Removal of Inhumanity* (*PHET, Chapter 4*). The core effort was about minimizing pain and distress in animals: less pain, better data. Ethics hardly came into it—perhaps a bit of human ethics, but certainly no animal ethics nor concern about insulting animals.

To be sure, I find the truism mentioned by Russell and Burch wholly inadequate as a reason to argue in favor of all future animal experimentation. It is a fundamental problem. It means that, from the outset, Russell and Burch and I are not on the same page. I strongly reject any speciesist stance that always, in every single case, places humans above other animals. Instead, I start from the explicit proposal that animal life deserves our compassion—a notion deduced from the meaning of *humane* (as listed in the Oxford English Dictionary). Importantly, I think animal life deserves our compassion—not because we are humans or because we wish to be considered humane, but because animal life has intrinsic value. Animals deserve our respect for their own sake.

We are free to do whatever we want, as long as we do no harm to others. This is the central principle in John Stuart Mill's classic defense of individual freedom, as he introduced it in *On Liberty* in 1859 (see a brief and very readable discussion in Sandel 2009, pp. 49–57). I think animals should be included among "others." I need to say more about why I think this, which I will do in Sect. 2.2 (*Changing Perspectives on the Use of Animals*). For now, I propose the following: We are free to do whatever we want, as long as we do no harm to others, including animals.

The first thing to acknowledge, then, is that any experiment that harms animals commits an ethical wrong. When is such a wrong justifiable? Which alternative concerns outweigh the harming of animals? I look for a reasonable argument based on unprejudiced scientific knowledge using internally coherent logic, which considers all parties concerned in the ethical questioning—including those that, by nature, cannot speak for themselves (just like we care for others who cannot speak for themselves, from babies to patients with Alzheimer disease). I insist we must provide a convincing rationale for each new animal experiment, to show how its promise of scientific or medical progress justifies the harming of animals. This involves an important balancing and a careful reasoning about welfare, both human and animal, which also takes opportunity costs into account (as I explain in Chapter 5). Before we start removing "inhumanity" in the technique, we would do well to question the objectives of each new animal experiment. Russell and Burch have nothing to say in this respect, which is why their approach to the three R's is in urgent need of revision; my task in this book is to specify which changes are required, both why and how. In the meantime, however, let us take a proper look at Russell and Burch's actual definitions of the three R's before attempting any redefinition.

2.1.3   The Actual Definitions

2.1.3.1 Replacement

The first, brief definition of Replacement is given in Chapter 4 of *PHET*, in the section that equated the removal of inhumanity with the three R's: "Replacement means the substitution for conscious living higher animals of insentient material." Chapter 5 of *PHET*, entirely dedicated to Replacement, opens with a fuller definition under the heading "Comparative Substitution":

> We shall use the term "replacement technique" for any scientific method employing non-sentient material which may in the history of experimentation replace methods which use conscious living vertebrates. Among this non-sentient material, we include higher plants, microorganisms, and the more degenerate metazoan endoparasites, in which nervous and sensory systems are almost atrophied.

Russell and Burch made a further distinction between Absolute and Relative Replacement (*PHET, Chapter 5, Modes of Absolute and Relative Replacement*):

> In relative replacement, animals are still required, though in actual experiment they are exposed, probably or certainly, to no distress at all. In absolute replacement, animals are not required at all at any stage.

Russell and Burch referred specifically to substitution by "insentient material." We note the categorical nature of the distinction: all or none, from "conscious living vertebrates" to material that has no sensory capacity. In Russell and Burch's view, replacing a monkey with a mouse would not constitute Replacement. Even replacing a monkey with a human fails their definition. However, if we choose to harvest a bit of neural tissue from a monkey before painlessly killing it, we might perform an in vitro experiment on non-sentient material—say, a patch-clamp intracellular recording of a hippocampal neuron, completely in accordance with the given definition of Replacement.

Tellingly, if somewhat inconsistently, Russell and Burch regarded Absolute Replacement as the "absolute ideal," implying (or implicitly acknowledging) that Relative Replacement did not quite guarantee the complete removal of distress. The definition also leaves invertebrates in an unclear category: unworthy of the same level of concern as conscious living vertebrates, but excluded from the range of alternatives.

It should be noted here that Russell and Burch clearly regarded Replacement as the preferable among the three R's, the one principle that "is always a satisfactory answer" (*PHET, Chapter 4, Contingent Inhumanity and the Problem of Scale*). Researchers should first consider Replacement and only move on to the other R's when they judge Replacement to be disadvantageous for the research. The priority of Replacement follows from the notion that it is the most efficient way to minimize distress. In line with the previous observations in *PHET*, the implication is that this focus on Replacement serves to promote science: less pain, better data. The surest way for this is to replace conscious living vertebrates by insentient material.

2.1.3.2 Reduction

In Chapter 4 of *PHET* (section *The Removal of Inhumanity: The Three R's*), the first, brief definition of Reduction states that, "Reduction means reduction in the numbers of animals used to obtain information of a given amount and precision." Here, the reduction is tied to the numbers of animals, not to the amount of distress. This definition of Reduction implies that a large amount of distress for a small number of animals is preferable over a small amount of distress for a large number of animals: The removal of "inhumanity" counts primarily the number of individuals involved.

Russell and Burch were careful to emphasize that their principle of Reduction is directly connected to the quality of research, particularly the reliability and reproducibility of findings. In Chapter 6 of *PHET*, devoted entirely to the topic of Reduction, they addressed the issue of sample size as follows (in a section with the title *The Problem of Variance*):

> In the real world, individual animals do vary. We can, therefore, never measure simply how animals of a given species respond to a given dose of a given substance. We have to take a *sample*, out of a population made up of all the other samples we could have taken at any time, and infer from the mean response of the sample chosen, combined with the variation within it, something about the effect of the treatment on any other sample we might have chosen. Our inference is of only relative accuracy, whose degree depends on the *size of the sample, the extent to which individuals of the species vary in response to the drug*, and *the efficiency in design and analysis of our experiment.* (italics in the original)

Russell and Burch added, in the subsequent section on *The Design and Analysis of Experiments*, "Every time any particle of statistical method is properly used, fewer animals are employed than would otherwise have been necessary." Statistical methods allow researchers to estimate the minimum number of animals necessary to be able to make valid inferences. The Reduction principle urges researchers to ensure they use *enough* animals, not less than the minimum number required for reliable conclusions. Russell and Burch regarded this as a matter of strategy and planning by individual researchers in their own laboratories. Given a research question, the principal investigator should calculate the minimum number of animals required to produce a reliable answer.

Critically, Russell and Burch did not formulate their principle of Reduction in absolute terms, but in relative terms. The research objectives determine the minimum number of animals needed in each case. Nothing is said about the number of research objectives (the number of research projects). This principle of Reduction, then, is entirely compatible with the possibility of large increases in the total number of animals used. If we have more questions, we will need to use more animals even if, for each particular question, we use the minimal number of animals required. As noted before, Russell and Burch never questioned the research objectives.

2.1.3.3 Refinement

The first, succinct definition of Refinement states that it "means any decrease in the incidence or severity of inhumane procedures applied to those animals which still have to be used" (*PHET, Chapter 4, The Removal of Inhumanity: The Three R's*). To the extent that animals are deemed necessary in research, the principle of Refinement requires researchers to use procedures that inflict less distress.

I have always found this the most confusing of the three R's; for students of my Bioethics class, it is invariably also the most difficult one to remember. The principle of Refinement sounds like a particular type of Reduction—not in the numbers of animals used, but in the amount of distress per individual animal. Given the goal of removing "inhumanity," it may seem to be a somewhat awkward choice for Russell and Burch to discuss the number of animals separately from the "severity" and "incidence" of procedures.

In fact, the amount of "inhumanity" turns out to be a slippery concept, in which we can distinguish an unfixed set of dimensions:

(a) The number of animals that have to suffer distress.
(b) The number of distressing procedures that animals have to suffer (which can be broken down further into the total number of distressing procedures, the average number of distressing procedures per animal, the maximum and minimum numbers of distressing procedures per animal, etc.).
(c) The severity of pain, or the level of distress in any given procedure, that animals have to suffer (which requires some kind of measurement or classification system, and again can be broken down further into the total severity in the applied procedures, or averaged, or maximum, minimum, etc.).
(d) The duration of the distress per individual procedure and across the animal's lifespan (where we might consider the number of procedures per time, but also the additional elements of distress, not due to specific procedures but due to the circumstances of life in captivity, including the types of housing, diet, etc.).

This is not a complete list. It is not obvious at all how we should address the removal of "inhumanity" in animal research through all these dimensions. We have no metric for the amount of distress. In this light, the suggestion that the three R's are developing a new science to remove "inhumanity" sounds rather premature. In Chapter 7 of *PHET*, Russell and Burch appeared to admit as much (in a section with the title *Neutral and Stressful Studies*):

Refinement, the third great path of advance, presents more formidable difficulties to the would-be taxonomist of techniques. It is indeed so protean in its aspects, that it would almost seem to require a separate solution in every single investigation, and refinement might be regarded as an art or an ability to improvise. It is true that the greatest experimenters have been artists in this sense, and that is one reason why we read with such aesthetic pleasure the accounts of their experiments.

I suspect art was on Russell and Burch's minds all along when they defined Refinement as the third R. In that fateful section of Chapter 4 in *PHET*, where they gave the first, brief definitions (*The Removal of Inhumanity: The Three R's*), the pair hinted at "the advantages of alliteration" and the usefulness of a "threefold structure" for mnemonic convenience. Given the natural two R's of Replacement and Reduction, Russell and Burch wanted to come up with a third. For want of a better word, they settled on Refinement as a second, fuzzier type of Reduction—not in the numbers of animals used, but in the amount of distress suffered by individual animals.

By simply rereading the original definitions, we can discover oddities, fuzziness, and gaps. Before continuing our critique of the three R's, however, I think we should reflect on the difference in context. Many good things (also many bad things) have happened in the last sixty years; the changes in context warrant an update of the three R's.

Some of the good things that have happened are due to Russell and Burch. I may complain of concept fatigue now, but the three R's deserve to be cherished as an inspirational point of departure for a more reasonable usage of animals in research. (I secretly believe Russell and Church would, if they were alive today, wholly endorse my critique of their work; I see my project as a Russell-and-Churchian effort with, ultimately, the exact same goal: a more reasonable usage of animals in research.)

2.2   Changing Perspectives on the Use of Animals

2.2.1   No Magic Dust

The Oxford Handbook of Animal Studies, as it was published recently (Kalof 2017), started its self-description on the inside flap of the dust jacket with the following observation:

Intellectual struggles with the "animal question"—how humans can rethink and reconfigure their relationships with other animals—first began to take hold in the 1970s. Over the next forty years, scholars from a wide range of fields would make sweeping reevaluations of the relationship between humans and other animals.

Russell and Burch (1959/1992) wrote their magnum opus decades before the sweeping reevaluations. We might be tempted to think they spoke well ahead of their time. More than fifteen years before the intellectual struggles began to take hold in the 1970s, they were already working in the laboratory to improve the treatment of animals in science. At the occasion of the fiftieth anniversary of the publication of *PHET*, Alan M. Goldberg and Michael Balls each gave a public lecture celebrating the important innovations with the three R's.

Goldberg (2010, p. 27) concluded, "*The Principles of Humane Experimental Technique*: Is It Relevant Today? The Answer: A resounding yes."

Balls (2010, p. 23) chimed in:

> The way in which Russell and Burch put it cannot be repeated too often: *If we are to use a criterion for choosing experiments, that of humanity is the best we could possibly invent. The greatest scientific experiments have always been the most humane and attractive, conveying a sense of beauty and elegance which is the essence of science at its most successful.*
>
> So, let us all take this opportunity to renew our commitment to live up to this ideal, with total sincerity, then go home, and get on with the job.
> (italics in the original)

I beg to differ. Rather than taking the opportunity to renew our commitment to live up to *that* ideal, I think we need a more fundament renewal, based on a different ideal. I certainly do not endorse simply getting on with business as usual. The key issue—the point I must keep harping on—is the misguided fixation on "humanity."

The desire for that ephemeral concept of humanity implies a categorical moral divide between *Homo sapiens* and the rest of the animal kingdom. It is a form of "human exceptionalism" (to borrow the label used by Driver 2011) that firmly belongs to the era before the sweeping reevaluations. Russell and Burch (1959/1992) did not escape the context of their time. They thought about animals and humanity in then-conventional ways.

Let us take another look at the inside flap of the dust jacket wrapped around our brand-new *The Oxford Handbook of Animal Studies*. The critical shift in perspective between the 1950s and today leaps up from two little words in the first paragraph, which twice uses the same little word: *other.* Let me use italics to signal the key formulation:

> Intellectual struggles with the "animal question"—how humans can rethink and reconfigure their relationships with *other* animals—first began to take hold in the 1970s. Over the next forty years, scholars from a wide range of fields would make sweeping reevaluations of the relationship between humans and *other* animals.

Humans are now firmly included in the category of animals. Not just our biological tissue bears great resemblance to that of other creatures. There is also the (somewhat unwoven, but still spellbinding) mystery of what goes on inside the meat between our ears. We are not as different from other creatures as we used to think. We have consistently been underestimating the nonhuman while overcomplicating the human (Rowlands and Monsó 2017). As it turns out, some nonhuman animal minds have far more capabilities than was known or acknowledged in mainstream science in the 1950s, the heyday of behaviorism. In the days when Russell and Burch crafted their masterpiece, experimental psychologists and neuroscientists (experts in the study of brains and behavior) generally espoused views that worked from the old Cartesian notion of animals as little robots composed of soulless mechanisms. More than half a century later, our handbook with the telling dust jacket devotes its entire second part (six chapters) to "Animal Intentionality, Agency, and Reflexive Thinking."

We used to think that humans are unique in their tool usage. Then one day, Jane Goodall observed a chimpanzee not just using a tool but making one, carefully selecting a twig and stripping the leaves so he (David Greybeard, as she called him) could use it to harvest termites. Goodall reported the data in *Nature* (1964) and went on to convince all kinds of audiences, and common sense, that what she had seen was no accident. (I had the good fortune of witnessing her remarkable narrative skills in Yokohama, in the summer of 2016.) In the meantime, we have learned that very many creatures can use tools. Hermit crabs acquire shells to reside in. They may even socialize to evict their neighbors (Laidre 2012). Ravens can plan flexibly for tool use, as well as for bartering, just like great apes (Kabadayi and Osvath 2017).

Upon first hearing of Greybeard's feat, the eminent paleontologist Louis Leaky famously replied, "Now we must redefine tool, redefine man, or accept chimpanzees as humans" (the quote lives on as a frequently-copied meme on the Internet; here, I am lifting it from the website of the Jane Goodall Institute Australia 2017). We did not redefine tool. Perhaps we tried to redefine what was special about humans, predominately by singing the praises of our faculty of language, but even there we ran into problems (e.g., Hauser et al. 2002; Traxler et al. 2012; Vicari and Adenzato 2014). In any case, I think most of us prefer to be careful about tying any special status of being human to linguistic abilities, lest we conclude that all creatures without the faculty of language should be regarded as subhuman. What would that imply for infants, the comatose, or anyone unable to pass a language test?

In an either/or scenario, either Descartes was wrong, and animals do have a soul. Or, Descartes was wrong, and humans do not have a soul. Thinking without overcomplicating, perhaps humans are *also* little robots composed of soulless mechanisms. They are just like other animals, although with more neural circuitry—a matter of degree. The idea seems less crazy, and is much better supported by science, than that of *Homo sapiens* sprinkled with a unique kind of magic dust (see Boden's monumental history of cognitive science with the title *Mind as Machine* 2006). In retrospect, we find that some of the great thinkers of the eighteenth and nineteenth centuries knew all along that human exceptionalism relied on fallacious thinking. In *A Treatise of Human Nature*, David Hume (1740/2002, p. 118) wrote in Sect. 1.3.16 (*Of the Reason of Animals*):

'Tis from the resemblance of the external actions of animals to those we ourselves perform, that we judge their internal likewise to resemble ours; and the same principle of reasoning, carry'd one step farther, will make us conclude that since our internal actions resemble each other, the causes, from which they are deriv'd, must also be resembling. When any hypothesis, therefore, is advanc'd to explain a mental operation, which is common to men and beasts, we must apply the same hypothesis to both; and as every true hypothesis will abide this trial, so I may venture to affirm, that no false one will ever be able to endure it.

Occam's razor applies. The same things are explained most straightforwardly in the same way. With Rowlands and Monsó (2017), we do well to reason in a minimalist vein about human and animal processes of the brain and mind. We can even cherish the old behaviorist approach while we are at it, as long as we apply the proper minimalist thinking equally to any phenomenon, for any creature. The same response to the same stimulus shall go by the same name, no matter whether the subject in our investigation is human or nonhuman—not a "feeling of fear" over here that is called a "conditioned reflex" over there.

Once we see through the fallacy of human exceptionalism, we must face the ethical implications. Jeremy Bentham suggested as much in a quote that I encountered twice (with interesting little variations) in *The Oxford Handbook of Animal Ethics* (2011, once in the essay by Nussbaum and once in that by Morris). Here I am quoting the fragment, via Google Books, without omissions or revisions—the proper classic (Bentham 1823, pp. 235–236):

> The day *may* come, when the rest of the animal creation may acquire those rights which never could have been witholden from them but by the hand of tyranny. The French have already discovered that the blackness of the skin is no reason why a human being should be abandoned without redress to the caprice of a tormentor. It may come one day to be recognized, that the number of the legs, the villosity of the skin, or the termination of the *os sacrum*, are reasons equally insufficient for abandoning a sensitive being to the same fate? What else is it that should trace the insuperable line? Is it the faculty of reason, or, perhaps, the faculty of discourse? But a full-grown horse or dog, is beyond comparison a more rational, as well as a more conversible animal, than an infant of a day, or a week, or even a month, old. But suppose the case were otherwise, what would it avail? the question is not, Can they *reason*? nor, Can they *talk*? but, Can they *suffer*? (italics in the original)

This certainly was futurist thinking in 1823. Bentham knew it. The day *may* come, as he italicized. Yet, the concept was there already. The key criterion was: a sensitive being. If the creature can suffer, it deserves some amount of care and protection. The rights would belong to the sensitive being itself, for no other reason than being what it is: a being that can suffer. Bentham anticipated a moral standing that did not depend on being human, by which a creature deserves protection against insult, not because such violation would (indirectly) diminish humanity, but because... the creature deserves protection. The buck would stop right there, with the creature.

2.2.2   For the Animal's Sake

What do most people think? The common view on the use of nonhuman animals in biomedical experimentation embraces two parts—on the one hand placing limits on what we can do to nonhuman animals, while on the other hand accepting some amount of usage when it significantly advances human interests (LaFollette 2011). This view emerges from a series of polls on animal rights conducted by the consulting company Gallup (Moore 2003; Newport 2008; Rifkin 2015). The polls were conducted always in the beginning of May, based on telephone interviews, with each sample consisting of more than a thousand adults (aged eighteen or over) living in the United States. Although the set of questions varied slightly over the years, the main item was asked every time:

Which of these statements comes closest to your view about the treatment of animals?

(a)   Animals deserve the exact same rights as people to be free from harm and exploitation.

(b)   Animals deserve some protection from harm and exploitation, but it is still appropriate to use them for the benefit of humans.

(c)   Animals don't need much protection from harm and exploitation since they are just animals.

Less than 4% of the respondents chose option *c* (consistently the same number, around 3%, in 2003, 2008, and 2015). The vast majority thinks animals deserve at lease some amount of protection. The most popular answer was *b* (71% in 2003, 72% in 2008, and 62% in 2015). The use of animals in research generally caused more concern than the welfare of animals in the other categories (animals kept in amusement parks, aquariums, or zoos, raised for food, or kept as household pets; as compared to animals in research, there were similar levels of concern for animals in the circus or in competitive sports). In 2015, more than two thirds of the respondents were at least somewhat concerned and one third were very concerned.

An interesting question, asked in 2003 and 2008 (but apparently not in 2015), was whether people supported the banning of all medical research on laboratory animals. A subtler line of questioning would have been preferable, but the results already suggested that people tend to think in less categorical terms. In 2003, the options were binary:

support or oppose. Oppose gained the upper hand (61%). This should not be taken to mean that the experimentation gets a free pass in people's minds. In 2008, the polling gave five options with respect to the banning of all medical research on laboratory animals (strongly support, somewhat support, somewhat oppose, strongly oppose, and no opinion). Only 33% of the respondents voiced strong opposition to *all* banning. Presumably, a good number of even these respondents might live with *some amount* of restriction.

LaFollette (2011) suggested that the common view on the moral permissibility of biomedical experimentation using animals sits somewhere near the center of a continuum that stretches from principally rejecting any usage of animals to the other extreme of categorically condoning everything. Comparing our current situation with that of the 1950s, we can safely conclude that the latter position, unconditionally promoting the use of animals in research, has lost terrain. One striking example is the banning or limiting of research using great apes, written in the law of the European Union and other countries (Project R&R 2017). The U.S. National Institutes of Health (NIH) announced not very long ago that it was ending its support for invasive research on chimpanzees (Kaiser 2015). Clearly, society as a whole is stepping away from an absolute moral divide between human and nonhuman animals. We are now moving on a slippery slope where the positions are unstable, shifting, and in need of careful and continued reassessment. We might recognize the changes as evidence of an expanding circle, an evolving morality (Singer 1981/2011).

Do nonhuman animals have moral status? Moral standing? Do our concerns pertain to the category of animal welfare or animal rights? This is certainly a dizzying topic, with a subtlety and variety of opinion I have yet to chart for myself, let alone for you, dear animal investor (here, I will simply refer to some of the scholars who have written on the topic: Beauchamp 2011; Copp 2011; Morris 2011). With Beauchamp, I agree that the distinction between animal rightists and animal welfarists is polarizing and must be rejected. Otherwise, we will get stuck. To move forward with improving the use of animals in research, we should focus on the commonalities in our views. I also like Beauchamp's phrasing of rights in terms of basic rather than "human" or "animal." As for whether nonhuman animals have moral standing, my answer depends on the definition.

I think animals exhibit various levels of sentience; some show evidence of types of cognition that has rational components in it. This is not to say those animals *are* rational. Even humans are not rational (only some humans some of the time; computers are better at being rational). I also think nonhuman animals are agents, but not moral agents. I do not think they ever act morally out of a moral sense. I have yet to see evidence of a nonhuman animal contemplating moral concepts the way we do (or the way we sometimes do, or rather too infrequently do). If sentience, rationality, *and* a moral sense are prerequisites to moral standing, then the available evidence tells me nonhuman animals do not have it.

By another definition, though, I do think nonhuman animals have moral standing (Andre and Velasquez 1991):

> What is moral standing? An individual has moral standing for us if we believe that it makes a difference, morally, how that individual is treated, apart from the effects it has on others. That is, an individual has moral standing for us if, when making moral decisions, we feel we ought to take that individual's welfare into account for the individual's own sake and not merely for our benefit or someone else's benefit.

When I consider using a monkey in research, I feel I ought to take that monkey's welfare into account—not for my sake, or for anybody else's sake, but for the monkey's sake. The monkey counts. Any intrusion on the monkey's welfare is morally wrong. I do not think it is necessarily, absolutely, and unconditionally wrong in every single case. Yet, it is a problem, a negative, a cost in the equation, an element in the reckoning that seems hard to avoid making, even if we do not like to call our reasoning utilitarian (we could give it another name; we could call it proportional thinking, something searching for a balance, a middle way, a golden mean, something Aristotelian, Confucian, or Buddhist; the name is of little concern to me, but the reasoning all the more). If I am to use the monkey, I had better have a very good reason for it—one that outweighs the monkey's welfare. I think this is not just my view. I take the data of the Gallup polls to mean that this is, in fact, the common view. Importantly, it is a view that clashes with the rationale underlying the three R's as proposed by Russell and Burch (1959/1992).

The objective of *PHET* was good science. The core hypothesis offered by Russell and Burch posited that the humane treatment of animals would actually benefit science. The removal of inhumanity—that is, the diminishing of animal distress—would be instrumental to the data extraction. At no point in their book did Russell and Burch suggest that the concern about causing animal distress in the pursuit of knowledge amounted to an ethical concern about committing something that was principally and morally wrong. They never stated it explicitly, but the underlying rationale for the three R's was that the ethical nature of the enterprise derived entirely from the notion that doing good science meant doing the right thing. Good science, or the most efficient pursuit of knowledge, was perfectly righteous, morally defensible, or even morally desirable. Anything that would advance the pursuit of knowledge was therefore ethically right—hence the desirability of removing animal distress not for the animal's sake, but for the sake of science and for the sake of humanity. This sounds about right in the context of the 1950s. Ultimately, this was an indirect form of animal ethics. The moral dimension belonged strictly to humans. To help humans and to gain more knowledge, Russell and Burch hypothesized it would be beneficial to remove animal distress. (The implication here is that, wherever science can be advanced by inflicting some amount of animal distress, the right thing to do is to go ahead and inflict the distress; there would never be any need to weigh costs versus benefits because the animal distress does not actually count as something morally wrong.)

In the Gallup polls, Russell and Burch would have sided with the 3%. They would have pointed out a hidden assumption in option *b* ("Animals deserve some protection from harm and exploitation, but it is still appropriate to use them for the benefit of humans"). This option works from the assumption that animal harm is the price for human benefit, exactly contrary to Russell and Burch's thesis. According to the inventors of the three R's, animal harm would also produce human harm (in the sense of poor science) and should therefore be avoided. In their ears, option *b* would have sounded nonsensical and unacceptable. For Russell and Burch, animal harm should never count as a direct concern in opposition to human benefit. The only logically coherent option, compatible with *PHET*, is *c* ("Animals don't need much protection from harm and exploitation since they are just animals")—the position today of a very small minority.

My reading of Russell and Burch's rationale neatly agrees with that of Tannenbaum and Bennett (2015, p. 123):

Although the fundamental aim of the 3Rs is diminishing or removing distress, for Russell and Burch this aim—and the use of the 3Rs—cannot be allowed to compromise the goals of conducting sound science and achieving scientific and medical progress. ... The goal of minimizing inhumanity is not to be balanced *against* the aims of a scientifically sound experiment or kind of research. (italics in the original)

Tannenbaum and Bennett (2015) wrote a careful analysis that compared the three R's as they were originally formulated against some of the later interpretations and applications. In all fairness, any satisfactory approach to the three R's should work from clear reasoning—either following the original definitions or supplying supporting arguments for revision. Tannenbaum and Bennett (p. 129) rightfully pointed out that, in many cases, the approach to the three R's departs from the original definitions without supplying a rationale:

Modified definitions of reduction typically are not accompanied by arguments supporting the view that it is better, irrespective of the effects on animal distress, to use fewer rather than more animals. It is incumbent on supporters of such definitions to provide such arguments.

On the same page, the pair of critics went on to guess at possible types of reasoning that might underlie the revisions:

Perhaps some proponents of such modified definitions of reduction believe that there is something inherently and unavoidably wrong in using animals in research—and that when it is necessary to use animals (a 'necessary evil' in such persons' view), using fewer animals is therefore better (a 'lesser evil') than is using more animals. Perhaps some proponents of such modified definitions of reduction would defend these definitions by pointing to potential savings in cost of research, difficulties and inconveniences in using animals in research, or opposition to animal use by some members of the public.

Tannenbaum and Bennett (2015), as we can taste from the distancing in the expressions ("such persons' view"), had little appreciation for the revisionists. The conclusion (on p. 131) left no doubt on the position of the authors: "Russell and Burch's original definitions, we submit, have much to recommend them." The definitions in *PHET* are entirely coherent and logical, from the overriding aim to conduct sound science, and from the subservient hypothesis that the removal of animal distress would benefit sound science.

If we are to depart from the original definitions, we do really have to clarify our position. I agree with Tannenbaum and Bennett's request, but not with their conclusion. *PHET* has had its moment. It is in urgent need of revision.

I offer a set of ten thoughts:

1. The aim to conduct sound science does not principally override the concern for animal welfare.
2. Russell and Burch's hypothesis about the relationship between animal distress and sound science is irrelevant and wrong. Sometimes, sound science causes animal distress. In such cases, the removal of animal distress diminishes sound science. In other words, sometimes the pursuit of sound science and the concern for animal welfare are indeed in opposition.
3. One key issue is animal distress.
4. Another key issue is sound science.
5. The key issues of animal distress and sound science are principally of concern to the entire community. The issues must be monitored and controlled through appropriate democratic mechanisms.
6. The approach to both issues requires careful consideration. Ultimately, it comes down to weighing different costs and benefits in each research project.
7. The weighing of costs and benefits should be done fairly, based on valid information, by the appropriate democratic authority.
8. Causing animal distress is morally wrong and an offense against the animal. It is intrinsically morally wrong for the animal's sake. Additionally, there are indirect types of moral issues associated— that is, offenses to certain humans or human groups, cultures, and institutions, among others.
9. We must try to assess the extent of animal distress as best we can, based on scientific knowledge.
10. When in doubt, we must let the concern for animal welfare take precedence over the pursuit of sound science. In general, the omission of sound science is morally less problematic than the commission of animal distress. Not gaining a bit of new knowledge should usually be considered preferable over causing animal distress. Only in clear-cut cases, when the prospective new knowledge is valued highly enough, can we consider the causing of animal distress justified.

Ultimately, the direct concern for animal welfare reflects the core difference between my position and that of Russell and Burch (1959/1992). Historically speaking, I think they were at the forefront of generating and promoting concern for animal welfare. Even though their rationale was based on an indirect concern (the true goal being sound science and humanity), in practice their proposal of the three R's effectively improved the use of nonhuman animals in research. Then, in the 1970s, the increased concern for animal welfare found a new theoretical grounding, perhaps most forcefully in Peter Singer's landmark publication of *Animal Liberation* (1975/2009). Singer focused on sentience; he argued that causing distress was morally wrong with respect to not only human beings, but also sentient nonhuman animals. Nonhuman animals are moral patients. They deserve our consideration for their own sake.

This was indeed a remarkable shift in thought: from indirect to direct concern for animal welfare. Critically, the shift in thought was congruent with science. It was based on the observation of similarity in structure and function between humans and other animals, with responses to stimuli in nonhuman animals that were recognized as indicative of pain and distress in human animals. Essentially, Singer (1975/2009) argued for an animal ethics that worked from scientific data, and he demanded a fair and coherent application of scientific knowledge in moral decision-making. Like should be treated alike. Ethical concerns with respect to distress should pertain to all subjects, not just the human. To the best of our scientific knowledge, there is no persuasive reason to consider human distress as categorically different from nonhuman distress solely on the basis of species membership. Indeed, based on the available data in neuroscience, ethology, biomedicine, and behavioral sciences, we cannot fail to recognize a continuum of behavioral, cognitive, and affective phenomena between human and nonhuman animals.

The idea of animal ethics was put on the map, and the promotion of animal welfare (or even animal rights) gained a noticeable amount of public support. In the present-day context, as the Gallup polls show, it has become a thing of common sense to assert that nonhuman animals deserve some amount of protection—a concern that weighs in opposition to human benefits. For the majority of people, the required protection of animals may be predominately a matter of moral intuition or unarticulated belief. These intuitions and beliefs run counter to conventional types of ethics that principally put humans first and see only indirect reasons to be concerned for animal welfare. For me, the required protection of animals is based on a comprehensive scientific view. I seek a coherent ethics—wishing to obtain what is good and avoid what is bad, based on valid information, with science and logic as the ultimate arbiters of what constitutes valid information. What is good and what is bad needs to be figured out, case by case, considering all the relevant data, including first and foremost what humans and other animals want or do not want.

The changed perspective on the use of animals, with the emergence of a direct concern for animal welfare, has led to a confusing array of differing interpretations of the three R's. We must reconsider our approach to the three R's, clear the rubble, and redefine the principles.

Before doing so, I would like to illustrate how the changed perspective on the use of animals throws new light on one particular type of research using one particular type of nonhuman animal—a type of research that was easily justified several decades ago, but not anymore (even researchers who still conduct this type of research realize they need to upgrade their defense; see Roelfsema and Treue 2014).

2.2.3   *The Monkey Eye-Movement Paradigm Then and Now*

The history of neuroscience is, to a significant degree, a history of animal experimentation. As one example of a line of research, I would like to focus on the study of the visual system, for the accidental reason that it happens to be the area I am most familiar with (see Lauwereyns 2012; a monograph on active vision). When reviewing the history of visual neuroscience, a good place to start is the work of David H. Hubel and Torsten N. Wiesel, who were jointly awarded one half of the Nobel Prize in Physiology or Medicine in 1981 (the other half going to Roger W. Sperry). Hubel and Wiesel received the ultimate honor in science "for their discoveries concerning information processing in the visual system" (Nobelprize.org 2017). The duo had, among other things, elucidated the way in which the physiological activity of retinal ganglion neurons (a type of neuron with its cell body in the retina and its axon leaving the eyeball, toward the brain) was based on opponent center versus surround processing, such that the neurons responded not so much to light or dark per se, but to *contrasts* between light and dark. I remember being utterly fascinated by the work when I first read Hubel's (1988) *Eye, Brain, and Vision*.

Hubel and Wiesel conducted their empirical research (from the mid 1950s to well into the present millennium) exclusively on animals, mostly cats and monkeys. Any textbook that refers to the neural mechanisms of visual processing essentially reaffirms the validity of this work. I am fully aware of the ethical implications whenever I teach a basic course on neuroscience and find myself showing the data to a new generation of students. In fact, I clearly mention it to students, without hesitation, without moral qualms: *This is data from cats.*

As a side note, I have seen people teaching similar materials while somehow dancing around the fact that it was based on animal research—as if there was some guilty conscience or ethical unease that made the truth unmentionable. In Chapter 5, I return to this theme of how to communicate about animal research, because it is a hugely important factor in the big picture of ethical reasoning. Ethical reasoning, I propose, must ultimately involve public engagement. Any teacher or researcher whose work benefits from animal research without acknowledging it does the research a great disservice. It may lead the public to underestimate the role of animal research in science. It may produce unfair evaluations of the costs and benefits. It gets even worse if teachers or researchers knowingly hide the fact that certain findings are based on animal research, simply because they are concerned about a potential negative public response. We do not need hypocrisy, but a fair and open assessment based on facts. If our work is somehow based on animal research, we have a moral duty to acknowledge that. If we disagree with the fact that the research was conducted on animals, we may add a note of dissent or append an opinion on how the progress might have been achieved differently. In any case, we must first realize how we got to where we are now, in order to assess how to move on in the best way possible (either in the same or in a different direction).

Banning hypocrisy, Hubel and Wiesel's legacy will remain untarnished. In using materials from their work for my teaching, I wholly endorse it. I think it represents a good example of animal research that was justified—in its time. The research did not quite do away with animal distress; the (undeniable) suffering of the cats here was clearly a cost that should be counted in opposition to the scientific value. Yet, in this case, retrospectively, I do believe the scientific value fully justified the cost. When looking back at the research, we must ask whether, by the standards of that time, the ethical rationale added up in favor of causing the animal distress. The standards of that time include not only the state of the art in science and the available research methods, but also the laws, research regulations, and the prevailing attitudes toward animals—not only in research, but also in the rest of society (including most conspicuously with respect to the food industry).

Let me illustrate the reasoning by considering another legacy that is very dear to me—that of Robert H. Wurtz ("Bob"). In the prelude to *Brain and the Gaze*, I wrote (2012, p. xx):

> If anyone has the appropriate expertise to write a book on brain and the gaze, it would have to be Bob Wurtz. Until he does finally write that book, I will merely stand in and write the present book in honor of the man who should have written it.

Although he initially performed experiments on rats (Wurtz and Olds 1963) and cats (e.g., Wurtz 1965), the mentor of my mentor came into his own as a scientist with research on the activity of visual neurons in monkeys that were making eye movements—a trilogy of publications in *Journal of Neurophysiology* around the time I was a newborn baby (Wurtz 1969a, b, c). The work represented a significant upgrade on the earlier research by Hubel and Wiesel, as it moved from studying passive vision (automatic responses to light) to studying active vision (natural behavioral dynamics while subjects were looking around). Vision, as it turned out, was not just a matter of linear processing from input to output, from stimulus to cognition or action. Vision proved to be deeply interactive and nonlinear—something that could not be captured by a conventional, simplistically deterministic structure. The work by Wurtz introduced a new paradigm, both theoretically and practically. Vision was itself sensorimotor (Wurtz's laboratory at the NIH would be dubbed the Laboratory of Sensorimotor Research). The way to study it was to have a monkey sit in front of a screen, wide awake, and make eye movements during visual tasks while the researcher lowered an electrode into the monkey's brain, lowering the tip deep enough, to very near the cell body of a single neuron, to record the so-called "spikes" (action potentials). For the next five decades, Wurtz and his colleagues and students harvested a great amount of data and a considerable body of knowledge about active vision using the monkey eye-movement paradigm.

If we turn back the clock to 1969, we might see a man on the moon, and me freshly delivered from the belly of one of that famous rebellious generation of the year before (although my mother was not actually among the fierce students who were then out in all force, inspired by their fellows in Paris, on the Bondgenotenlaan in Leuven, the main boulevard of the city of my alma mater). It was a handful of years after Bob Dylan had started singing of his well-known hypothesis of a paradigm shift. Meanwhile, the other Bob—Bob Wurtz—was busy in the laboratory, charting the receptive fields of striate cortex neurons in monkeys as they were making eye movements. What can we say about the context?

In the great pursuit of knowledge, it was the early days for neuroscience. There were no computers in the laboratory, no tools for neuroimaging in humans (no scanners to trace cerebral blood flow), and hardly any data about the functional activity of living brains. Behaviorism still ruled the airwaves in psychology, compatible with the Cartesian view of animals as mere robots. The methods available to neurophysiology were crude, imprecise, and manageable only at a certain scale (scientists liked the axons of the giant squid for being so big and unmyelinated, and thus easy to prick an electrode in). The monkey brain was preferable over that of a rat or a mouse—not only for its relative similarity to the human brain, but also for its size, which was large enough to perform the not-so-delicate operations on.

The context in society, more generally, had begun changing, but the changes were still few. In Western countries, white men ruled. There was a large gap between laypeople and experts (no students could torture their professors about facts they had instantly, if mindlessly, retrieved from Wikipedia). Patriarchal, autocratic forms of leadership were standard in most professions, and certainly in the medical sector. The concept of ethics did not seem to apply to animals.

In those days, considering what was known and what was possible in science, as well as what was happening in the rest of society, Wurtz had every right to perform his experiments on monkeys. It was a matter of free inquiry, affording a large gain of knowledge in a short amount of time. The experiments were justifiable. In fact, nobody stood up to question them. Moreover, Wurtz, being the conscientious researcher, duly applied the three R's—or at least one of the three, Refinement. He anesthetized his monkeys before removing their scalps and opening their skulls. Indeed, the anesthetics were helpful in getting the subjects to stay very quiet throughout the process. After they recovered from surgery, Wurtz treated the monkeys well enough so they were willing to come out of their cages to sit in the Plexigas chair, get hooked up to the machine, and earn a bit of water or juice for making eye movements to dots on the screen.

Three decades later, after a Ph.D. in cognitive psychology, I joined the laboratory of one of Wurtz's most talented disciples, Okihide Hikosaka, who was then at Juntendo University in Tokyo, Japan, and who would eventually replace Wurtz in the Laboratory of Sensorimotor Research at the NIH. We conducted research on motivational aspects of the oculo-motor system, achieving a fine set of research outputs with essentially an unchanged paradigm (e.g., Lauwereyns et al. 2002a, b). The only mod-ifications pertained to the degree of sophistication in the task design and data analysis, aided by better tools (computers) and new concepts (new to the field of neuroscience, but borrowed from decades-old work in cognitive psychology).

Was this work justifiable, ethically? I entered the research field with something like a "suspension of disbelief" (to appropriate the famous concept by the poet Samuel Taylor Coleridge); I had my doubts, but I was no expert on the matter and no animal rights activist. I could not presume to know better than the researchers. I was (and still am) sim-ply, genuinely, interested in cognitive neuroscience. I just wanted to know. Gradually, however, my doubts turned into serious skepticism as I noticed gaps and inconsistencies in the ethical reasoning. In Chapter 4, I address my misgivings in detail.

Here, I merely note that my worries about the ethical reasoning did not emerge in isolation. Something had changed—not with the monkey eye-movement paradigm, but with its reception. It no longer got a free pass from the public. I remember, at the turn of the millennium, reading panicky editorials in *Nature Neuroscience* decrying terrorist attacks on scientists (1999), legal challenges to animal experimentation (2000), and the threat of extinction for several primate species (2001; the editorial appeared to lament particularly the disappearance of research opportu-nities). It sometimes looked as if all of society was against "us." Yet, my question was not, "How then should neuroscientists defend their work?" It was broader: "Where did we go wrong?" Perhaps, as implied by the editorials, the problem was one of public engagement—a lack of com-munication by scientists. That could be true. Additionally, there could be something more fundamental going on—a problem with the work itself. Perhaps the research itself needed to be questioned—not just by the ("ignorant") public, but also by informed scientists.

Today, another two decades later, we are still grappling with the old definitions of the three R's and continuing the monkey eye-movement paradigm—still essentially unchanged, but now with even more powerful computers and even more sophisticated task designs (e.g., Barack et al. 2017; Kim et al. 2017; White et al. 2017). Is this work justifiable, ethically? Considering what is known, what is possible in science, and what is happening in the rest of society, the picture is certainly cloudier now than in the early days for Wurtz.

Neuroscience has achieved enormous progress in the past five decades. Behaviorism has become a bit player. Even in hardcore neuroscience, we now openly study the once-taboo phenomena of emotions, feelings, and consciousness. Based on solid scientific evidence, we have come to appreciate the cognitive abilities of certain (nonhuman) animals. If *they* are robots, *we* are robots. If we refuse to be viewed as robots, we must refuse to view them as robots too. Thanks to powerful computers and scanners, we now have good, noninvasive techniques for neuroimaging in humans. The work with rodents (e.g., using optogenetic tools) has not simply rivaled, but has completely overtaken, the work with primates at the cutting edge—the true frontier of innovative neuroscience. By any objective measure, the role of primates in this branch of the pursuit of knowledge has diminished vastly.

Meanwhile, the changes in society more generally have gained speed. In Western countries, white men no longer rule by default. The first black president of the United States of America is no longer a distant fantasy, but (for the majority of citizens in that country, and for an even bigger majority of citizens outside that country) a cherished memory, regretted only for the fact that it is in the past. Angela Merkel serves as Chancellor of Germany; she has been doing so for over ten years, valiantly defending democratic principles. Patriarchal, autocratic leadership does not match with the tenets of Western societies. Racism and sexism are vigorously challenged around the globe—despite, or in defiance of, violent regressive movements. The Internet and social media give easy access to information. We can analyze big data. Economic pressures emphasize the need for wise investment. The concept of animal ethics has come into being, along with notions such as animal rights and animal rights activists.

The eye-movement paradigm with monkeys no longer gives a large gain of knowledge in a short amount of time, especially in comparison to the alternatives available. We are facing diminishing returns and higher costs. The work seems increasingly more difficult to justify. People are standing up to question it. Yet, the three R's do not help us on our way to a reassessment. Researchers who are stuck in the eye-movement paradigm with monkeys happily wave in the direction of the three R's (Roelfsema and Treue 2014). The principles are on their side. There is no nonsentient tissue that can replace an eye-moving monkey, and the researchers are doing the best they can in working with the original definitions of Reduction and Refinement.

Something must be wrong with the three R's. They are no longer suited to the present context. Before jumping to a new set of definitions for the three R's, I think we must get a clearer view on how researchers can rightfully claim to be using Russell and Burch's (1959/1992) principles while at the same time effectively contributing to their failure. This requires an analysis inspired by studies in economics and social sciences, comparing individual agency with group dynamics: my topic for Chapter 3.

References

Andre, C., & Velasquez, M. (1991). Who counts? *Issues in Ethics, 4*(1). Available at: https://legacy.scu.edu/ethics/publications/iie/v4n1/counts.html.

Balls, M. (2010). The principles of humane experimental technique: Timeless insights and unheeded warnings [Special Issue]. *ALTEX, 27,* 19–23.

Barack, D. L., Chang, S. W. C., & Platt, M. L. (2017). Posterior cingulate neurons dynamically signal decisions to disengage during foraging. *Neuron, 96,* 339–347.

Beauchamp, T. L. (2011). Rights theory and animal rights. In T. L. Beauchamp & R. G. Frey (Eds.), *The Oxford Handbook of Animal Ethics* (pp. 198–227). New York: Oxford University Press.

Bentham, J. (1823). *An Introduction to the Principles of Morals and Legislation.* A new edition, corrected by the author (Vol. II). London: W. Pickering.

Boden, M. A. (2006). *Mind as Machine. A History of Cognitive Science* (Vols. 1 and 2). New York: Oxford University Press.

Buchanan-Smith, H. M., Rennie, A. E., Vitale, A., Pollo, S., Prescott, M. J., & Morton, D. B. (2005). Harmonising the definition of refinement. *Animal Welfare, 14,* 379–384.

Copp, D. (2011). Animals, fundamental moral standing, and speciesism. In T. L. Beauchamp & R. G. Frey (Eds.), *The Oxford Handbook of Animal Ethics* (pp. 276–303). New York: Oxford University Press.

Cornwall, W. (2017). Revamp animal research rules, report urges. *Science, 358,* 434.

Driver, J. (2011). A Humean account of the status and character of animals. In T. L. Beauchamp & R. G. Frey (Eds.), *The Oxford Handbook of Animal Ethics* (pp. 144–171). New York: Oxford University Press.

Editorial. (1999). Science and terrorism in Europe. *Nature Neuroscience, 2,* 99–100.

Editorial. (2000). Legal challenges to animal experimentation. *Nature Neuroscience, 3,* 523.

Editorial. (2001). Primate research faces extinction. *Nature Neuroscience, 4,* 111.

Goldberg, A. M. (2010). The principles of Humane experimental technique: Is it relevant today? [Special Issue]. *ALTEX, 27,* 25–27.

Goodall, J. (1964). Tool-using and aimed throwing in a community of free-living chimpanzees. *Nature, 201,* 1264–1266.

Hauser, M. D., Chomsky, N., & Fitch, W. T. (2002). The faculty of language: What is it, who has it, and how did it evolve? *Science, 298,* 1569–1579.

Hubel, D. H. (1988). *Eye, Brain, and Vision.* Scientific American Library. New York: W. H. Freeman.

Hume, D. (1740/2002). *A Treatise of Human Nature* (D. F. Norton & M. J. Norton, Eds.). New York: Oxford University Press.

Kabadayi, C., & Osvath, M. (2017). Ravens parallel great apes in flexible planning for tool-use and bartering. *Science, 357,* 202–204.

Kaiser, J. (2015, November 18). NIH to end all support for chimpanzee research. *Science,* News online. https://doi.org/10.1126/science.aad7458.

Kalof, L. (Ed.). (2017). *The Oxford Handbook of Animal Studies.* New York: Oxford University Press.

Kim, H. F., Amita, H., & Hikosaka, O. (2017). Indirect pathway of caudal basal ganglia for rejection of valueless visual objects. *Neuron, 94,* 920–930.

LaFollette, H. (2011). Animal experimentation in biomedical research. In T. L. Beauchamp & R. G. Frey (Eds.), *The Oxford Handbook of Animal Ethics* (pp. 796–825). New York: Oxford University Press.

Laidre, M. E. (2012). Niche construction drives social dependence in hermit crabs. *Current Biology, 22,* R861–R863.

Lauwereyns, J. (2012). *Brain and the Gaze: On the Active Boundaries of Vision.* Cambridge, MA: The MIT Press.

Lauwereyns, J., Takikawa, Y., Kawagoe, R., Kobayashi, S., Koizumi, M., Coe, B., et al. (2002a). Feature-based anticipation of cues that predict reward in monkey caudate nucleus. *Neuron, 33,* 463–473.

Lauwereyns, J., Watanabe, K., Coe, B., & Hikosaka, O. (2002b). A neural correlate of response bias in monkey caudate nucleus. *Nature, 418,* 413–417.

Moore, D. W. (2003, May 21). Public lukewarm on animal rights. *GALLUP News.* Available at: http://news.gallup.com/poll/8461/Public-Lukewarm-Animal-Rights.aspx.

Morris, C. W. (2011). The idea of moral standing. In T. L. Beauchamp & R. G. Frey (Eds.), *The Oxford Handbook of Animal Ethics* (pp. 255–275). New York: Oxford University Press.

Newport, F. (2008, May 15). Post-derby tragedy, 38% support banning animal racing. *GALLUP News.* Available at: http://news.gallup.com/poll/107293/PostDerby-Tragedy-38-Support-Banning-Animal-Racing.aspx.

Nobelprize.org. (2017, December 5). The Nobel prize in physiology or medicine 1981. Nobel Media A B 2014. Available at: https://www.nobelprize.org/nobel_prizes/medicine/laureates/1981/index.html.

Nussbaum, M. (2011). The capabilities approach and animal entitlements. In T. L. Beauchamp & R. G. Frey (Eds.), *The Oxford Handbook of Animal Ethics* (pp. 228–251). New York: Oxford University Press.

Olsson, I. A. S., Franco, N. H., Weary, D. M., & Sandøe, P. (2012). The 3Rs principle: Mind the ethical gap! *ALTEX Proceedings, 1/12, Proceedings of WC8, 29,* 333–336.

Project R&R. (2017). *International bans: Countries banning or limiting chimpanzee research.* Available at: http://www.releasechimps.org/laws/international-bans.

Rifkin, R. (2015, May 18). In U.S., more say animals should have same rights as people. *GALLUP News: Social Issues.* Available at: http://news.gallup.com/poll/183275/say-animals-rights-people.aspx.

Roelfsema, P. R., & Treue, S. (2014). Basic neuroscience research with nonhuman primates: A small but indispensable component of biomedical research. *Neuron, 82,* 1200–1204.

Rowlands, M. J., & Monsó, S. (2017). Animals as reflexive thinkers: The aponoian paradigm. In L. Kalof (Ed.), *The Oxford Handbook of Animal Studies* (pp. 319–341). New York: Oxford University Press.

Russell, W. M. S., & Burch, R. L. (1959/1992). *The Principles of Humane Experimental Technique.* Wheathampstead: Universities Federation for Animal Welfare. Available at ALTWEB: http://altweb.jhsph.edu/pubs/books/humane_exp/foreword.

Sandel, M. J. (2009). *Justice: What's the Right Thing to Do.* New York: Farrar, Straus and Giroux.

Singer, P. (1975/2009). *Animal Liberation.* New York: HarperCollins.

Singer, P. (1981/2011). *The Expanding Circle: Ethics, Evolution, and Moral Progress.* Princeton, NJ: Princeton University Press.

Tannenbaum, J., & Bennett, B. J. (2015). Russell and Burch's 3Rs then and now: The need for clarity in definition and purpose. *Journal of the American Association for Laboratory Animal Science, 54,* 120–132.

The Jane Goodall Institute Australia. (2017). *About Dr. Jane Goodall.* Available at: http://www.janegoodall.org.au/about-dr-jane-goodall.

Traxler, M. J., Boudewyn, M., & Loudermilk, J. (2012). What's special about human language? The contents of the "narrow language faculty" revisited. *Language and Linguistics Compass, 6,* 611–621.

Vicari, G., & Adenzato, M. (2014). Is recursion language-specific? Evidence of recursive mechanisms in the structure of intentional action. *Consciousness and Cognition, 26,* 169–188.

White, B. J., Kan, J. Y., Levy, R., Itti, L., & Munoz, D. P. (2017). Superior colliculus encodes visual saliency before the primary visual cortex. *Proceedings of the National Academy of Sciences of the United States of America, 114,* 9451–9456.

Wrangham, R. (2004). Killer species. *Daedalus, 133,* 25–35.

Wurtz, R. H. (1965). Steady potential shifts during arousal and deep sleep in the cat. *Electroencephalography and Clinical Neurophysiology, 18,* 649–662.

Wurtz, R. H. (1969a). Comparison of effects of eye movements and stimulus movements on striate cortex neurons of the monkey. *Journal of Neurophysiology, 32,* 987–994.

Wurtz, R. H. (1969b). Response of striate cortex neurons to stimuli during rapid eye movements in the monkey. *Journal of Neurophysiology, 32,* 975–986.

Wurtz, R. H. (1969c). Visual receptive fields of striate cortex neurons in awake monkeys. *Journal of Neurophysiology, 32,* 975–986.

Wurtz, R. H., & Olds, J. (1963). Amygdaloid stimulation and operant conditioning in the rat. *Journal of Comparative and Physiological Psychology, 56,* 941–949.

CHAPTER 3

A Mismatch Between Micro-motives and Macro-behavior

Abstract Viewed at the microscopic level of each individual laboratory, it may appear that researchers work in accordance with the three R's, choosing the appropriate animal model, aiming to get the smallest sample size needed for any experiment, and continuing to improve the techniques and knowledge extraction. However, turning to the macroscopic level of the entire field, we find mismatches between the individual intentions and overall outcomes. The chapter offers an analysis inspired by the groundbreaking work of Schelling on emergent patterns of macro-behavior as a function of micro-motives. The challenge will be to shift from agency at the microscopic level to agency at the macroscopic level. Such a shift would enable researchers to address, among other things, the key scientific problem of sample size.

Keywords Animal ethics · Rational choice · Individual perspective
Aggregate outcome · Sample size

The number of animals used in research worldwide is difficult to estimate because many countries either do not collect such data or report the data only partially. In the United Kingdom, the number of procedures using nonhuman vertebrate animals peaked in the mid 1970s (well above 5 million experiments annually) and then declined to nearly half that number in the late 1990s; since then, their use has been increasing again to about 4 million experiments annually in recent years (based on data up

© The Author(s) 2018
J. Lauwereyns, *Rethinking the Three R's in Animal Research*,
https://doi.org/10.1007/978-3-319-89300-6_3

to 2016, as collated by Understanding Animal Research 2017). Mice are the largest group (72.8% in 2016), with comparatively small numbers of procedures on cats, dogs, horses, and primates (the latter group amounted to 0.09% of the total, or 3569 procedures in Great Britain).

Even the 2016 numbers from the United Kingdom are difficult to compare directly with the United States. The United states Department of Agriculture (USDA) counts animals, not procedures; it also does not track the numbers of rats, mice, fish, and birds used in research. The 2016 report (United States Department of Agriculture 2017) listed more than 800,000 animals. Assuming the ratios are comparable to those in the United Kingdom, we may infer that the total number including rats, mice, fish, and birds exceeds 20 million animals. Interestingly, there appears to be a relatively higher number of nonhuman primates in research in the United States than in the United Kingdom; more nonhuman primates than dogs and horses are used in the United States, whereas the reverse is true in the United Kingdom. The USDA report gave a total number of 71,188 nonhuman primates used (collated across "all pain types"; 1272 were used in research that involved pain, without drugs to relieve the pain).

What are we to make of these numbers? In one sense, they pale in comparison to the numbers of animals used in the food industry, where there were "9 billion broiler chickens alone" (Rollin 2017, p. 345). Roelfsema and Treue (2014, p. 1200) computed the following:

> In European countries for every single research animal, about 200–300 animals are killed for human consumption. Of the research animals used, more than 80% are rodents and less than 0.1% are nonhuman primates Thus, we consume about 500,000 animals for every nonhuman primate in research.

Written between the lines here seems to be that we should not get too upset about that one monkey. Roelfsema and Treue did not suggest that humans should reduce their level of animal consumption. Looked at from another angle, the numbers—dazzling though they are—seem somewhat misleading by mixing categories. How many nonhuman primates are eaten in European countries every year? I do not know where to find data on the consumption; however, I think it is fair to estimate that the number of eaten monkeys is low—certainly lower than the number used in research.

On another occasion, we might review the issues with respect to the consumption of nonhuman primates as bush meat in countries outside Europe. (To be sure, I object to any such consumption, but the topic is beyond our scope here.) I do agree with Roelfsema and Treue's implication that killing animals for research is not necessarily worse than killing animals for consumption. Yet, one wrong over here does not excuse another wrong over there. The wrongs should be stopped wherever we can. In our use of animals, I would like to move toward a coherent set of principles; in this sense, any advancement made with respect to the use of animals in research can lead the way for other areas of society.

Roelfsema and Treue's sleight of hand in their computation—mixing categories—is completely compatible with Russell and Burch's (1959/1992) definition of Replacement. In that view, there are really only two categories: sentient versus non-sentient, meaning that monkeys are indeed in the same category as pigs, chickens, and mice. I think we need to use a more sophisticated categorization, with the help of science and with a better understanding of the different levels and types of sentience. This monograph is working toward that redefinition. At this point in the discourse, I note a request: The principle should provide consistency across different kinds of animal use, not only for research.

The paragraph immediately following Roelfsema and Treue's numerical estimate (2014, still on p. 1200) is worth reproducing in full because it exemplifies several stereotypes that are used to defend the status quo. I use italics to signal the revealing elements:

> Because of the *broad public agreement* that *some animal research* is necessary to ensure human health and medical progress, *groups opposed to any animal research* have refocused their *broad assault* onto just "basic" (as opposed to "applied") research and on nonhuman primates (as opposed to the vast majority of other species used). We will therefore focus here on basic neuroscience research with monkeys (nonhuman primates [NHPs]) as a relatively small but essential part of biomedical research. It has provided the basis for groundbreaking discoveries and progress but has also been the focus of *very vocal and sometimes violent opposition* from *well-funded groups* waging a campaign against animal-based biomedical research.

The "broad public agreement" may be less broad than casually assumed here. In Chapter 2, I argued how the societal context has changed considerably since the 1970s. A series of Gallup polls tracking the change over time in the moral acceptability of twenty issues found that medical testing on animals was considered acceptable by 56% of the American public in 2013, down from 65% in 2001 (Newport and Himelfarb 2013). The majority said "yes," but the majority was getting smaller—too small to be considered "broad." In fact, the downward trend was the largest of all twenty issues; if the trend continues in the same way, the next Gallup poll might find that the majority of the general public opposes medical testing on animals.

The public would agree with "some animal research," not specifying any species—"some," that is, not all. How much public agreement would there be for medical testing on monkeys, non-medical testing on monkeys, and taxpayer-funded non-medical testing on monkeys? I predict the numbers would fall well below 50% if Gallup asked members of the general public, "Would you be willing to pay for scientific experiments on monkeys?" It is not wise, in fact it may be wishful thinking, to assume the public broadly agrees with a status quo in the use of all types of animals for research.

Roelfsema and Treue (2014) were motivated to write their defense against a perceived "broad assault" by "groups opposed to any animal research." This opposition is further characterized as "very vocal and sometimes violent" at the end of the cited paragraph. The language reinforces a polarization, of "us" being assaulted by a violent "them"—a fight in which scientists are seen as a homogeneous group, with broad public agreement on their side. Curiously, Roelfsema and Treue alluded to the opposition groups as being "well-funded," without specifying which was the generous source of funding (the paragraph did not refer to any verifiable data). It might have been crowd-funding and charitable contributions—from the same general public that was supposed to be on the scientists' side.

Roelfsema and Treue's (2014) defense of basic neuroscience research using nonhuman primates appeared in the specialist journal *Neuron*—a journal that regularly publishes such research. Why put out a defense there? Most of the article consisted of an enumeration of past achievements with nonhuman primate research. Does the audience of *Neuron* need a reminder that the work with nonhuman primates is really "necessary"? (In Chapter 4, I deconstruct this concept of necessity.) Perhaps the authors wanted to rally support and give their fellow researchers a manual on what to communicate to the general public. The last section of the article is titled, "Informing the Public about the Necessity of Research Involving NHPs and the Efforts to Minimize Harm" (p. 1203). The latter part of the title gives a nod to the three R's, particularly Refinement.

The article illustrates how the three R's have lost their power to improve the rationale and practice of animal research. They have become a handy ethical disclaimer utilized in a defense of the status quo. Yet, for the past two decades, the numbers of animals used in research have been on the rise again; at the same time, public support for animal research has shown a downward trend. Roelfsema and Treue—two senior, very successful scientists—can work with the prevailing ethical principles to argue that nothing should change, but the very fact they feel compelled to write a defense already signals that all is not well.

I believe insisting on the status quo is wrong. The polarizing strategy is misleading, and researchers and scholars do not form a homogeneous group. Some are skeptical about the benefits and the moral acceptability of certain types of animal research, but not all critics of animal research are violent or irrational. The polarizing strategy—aiming for a tug of war between "us" and "them"—could be counterproductive, possibly alienating the general public. Instead of polarizing, I suggest that cooperating with the general public (including skeptical researchers and scholars from different fields) is key to a continued, healthy amount of support for research. For this, researchers would do well to heed cues from the general public, rethinking how and when to use animals for research, as well as how to share knowledge about it through outreach.

In Chapter 2, I argued that the original definitions of the three R's are out of touch with contemporary ideas about animals, what we know about them, and how we think we should treat them. Here, I focus on inherent conceptual weaknesses with the three R's, effectively creating confusion and offering researchers something like a loophole—a free pass to stop worrying about ethics and carry on working as before. As a result, the way animal research is conducted today actually violates the basic gist of the three R's. Researchers, while complying with the three R's and rightfully getting permission from animal ethics committees, do not succeed in removing animal suffering. To understand this apparent paradox, we must analyze the way we use the three R's, at the individual level versus at the group level; this means analyzing aggregate behavior.

3.1   ANALYZING AGGREGATE BEHAVIOR

In social sciences, and especially in economics, we have known for a good while that human behavior can be analyzed at different scales, from micro to macro. We can consider the behavior of individual consumers, zooming in on specific intentions, or we can zoom out and view entire markets to track the dynamics of supply and demand locally, nationally, or internationally. One prominent lesson from the analysis at different scales is that the best intentions of individuals do not necessarily—in fact, often do not—add up to the optimal solution for those same individuals. The whole can end up being very different from the sum of its parts. Famous examples include the prisoner's dilemma (see Kuhn 2017, for a comprehensive introduction) and the tragedy of the commons (Hardin 1968). Both paradigms have in common that individuals, making rational decisions, end up contributing to a suboptimal endpoint for everyone. These paradigms show mismatches between "micro-motives and macro-behavior," to borrow the expression by Thomas C. Schelling (1978/2006; possibly the economist most associated with this kind of analysis, for which he was awarded the 2005 Nobel Memorial Prize in Economic Sciences).

I think the implementation of the three R's in animal research also exhibits a mismatch between micro-motives and macro-behavior. I do not wish to presume that everyone is deeply familiar with the classic examples of this phenomenon. To be able to show how the micro-macro analysis is relevant here, I will briefly introduce some examples—one in my own wording and two in the unbeatable wording of the thinkers who first thought them (readers who are in a rush and already know the stuff by heart can skip ahead to Sect. 3.1.2).

3.1.1   Three Classic Examples of Micro-Macro Conflict

3.1.1.1   The Prisoner's Dilemma

Imagine you and I have robbed a bank—and we nearly got away with it. We were able to hide the two hundred thousand dollars, but the police captured us in the end. They have no evidence, though. All they can prove is that we were driving a stolen car. For that, they can put us in jail—you one year, me one year. However, they consider us the prime suspects for the bank robbery; they just need one of us to confess and make a deal with them. If you confess and I stay silent, I get the blame, you go free, and I get three years in jail (vice versa if I confess and you stay silent). However, if we both confess, all deals are off; we share the blame, and we both get two years in jail.

This is the prisoner's dilemma. The optimal solution for us both—the rational aggregate behavior—requires that you and I stay silent. Unfortunately, we will fail to do so. Each of us, thinking rationally from our individual perspective, cannot escape the conclusion that we should confess. If I stay silent, chances are that I get one year in jail (if you stay silent too) or that I get three years in jail (if you confess). One or three years comes to an average expectancy of two years. The picture looks better for me if I confess. In that case, chances are that I get two years in jail (if you also confess) or that I go free (if you stay silent). Two or zero years gives an average expectancy of one year. Thus, logic demands that I confess. Logic similarly demands that you confess. Logic will drive us away from the optimal solution. We will end up at the suboptimal equilibrium—the situation in which we both have to serve two years in prison.

Of course, in real life, we can think of situations in which people do go beyond the boundaries of individual logic. If there is something that binds you and me with unshakable trust and utter conviction, or if we fear some form of punishment after betrayal, we will manage to cooperate. I will trust you to stay silent, and I will stay silent, too. We will achieve that distant goal—it is not impossible. In fact, humans manage to cooperate to an amazing extent compared with other animals, thanks to various social mechanisms (e.g., reputation making, punishment of defectors, punishment of those who fail to punish defectors; see Fehr and Fischbacher 2003).

3.1.1.2 The Tragedy of the Commons

The second classic example I would like to raise is the tragedy of the commons, as introduced by Hardin (1968, p. 1244; italics in the original):

> Picture a pasture open to all. It is to be expected that each herdsman will try to keep as many cattle as possible on the commons. (...) As a rational being, each herdsman seeks to maximize his gain. Explicitly or implicitly, more or less consciously, he asks, "What is the utility *to me* of adding one more animal to my herd?" (...) Adding together the component partial

utilities, the rational herdsman concludes that the only sensible course for him to pursue is to add another animal to his herd. And another; and another.... But this is the conclusion reached by each and every rational herdsman sharing a commons. Therein is the tragedy. Each man is locked into a system that compels him to increase his herd without limit—in a world that is limited.

Hardin neatly identified the core problem, the clashing of limits. Any shared resource in the real world is likely to be limited—renewable perhaps, but limited. We can think of fossil fuels, forestry, fisheries, and any number of examples. There is only so much of the resource, and over-usage may deplete or destroy it. The limits of the shared resource determine the limits of healthy aggregate behavior. These macroscopic limits are typically beyond the horizon from the microscopic perspective, invisible to the individual. Seeing only what is close by, we are more likely to be concerned with our freedom to choose what is best for us. Yet, freedom and rationality, organized at the microscopic level, spell bad news for us all. Hardin did not mince his words. Immediately following the fragment cited above, he concluded:

Ruin is the destination toward which all men rush, each pursuing his own best interest in a society that believes in the freedom of the commons. Freedom in a commons brings ruin to all.

Instead of freedom, Hardin (1968) argued for the recognition of necessity, studying the patterns at the macroscopic level and drawing conclusions about how to limit the individual's freedom in a way that is compatible with the limits of the resource in question. Thus, we would manage "mutual coercion mutually agreed upon" (p. 1247), a higher-order freedom, in which we freely choose to moderate our intake.

In recent empirical work on the tragedy of the commons, using a so-called intergenerational goods game, Hauser and colleagues (2014) found that perhaps the picture was slightly less bleak than what Hardin (1968) had suggested. Not everyone chooses relentlessly for their own benefit, regardless of the potential cost to the group. However, the damage is ultimately the same—*done*—if only a minority of individuals cannot help themselves from feeding excessively on the dwindling. To cooperate with the future, we need to keep each other in check. By applying the good old principles of democracy, a majority of cooperators can restrain a minority of defectors. It also helps people who are sitting on the fence to choose to cooperate rather than to defect. Yet, the rules have to apply to all; furthermore, without sanctions for betrayal, the rules have no meaning. The power—almost tautologically—is in the reinforcement, applying the proverbial carrot and stick.

3.1.1.3 The Case of the Two Dining Rooms

My third example of micro-macro analysis (also a classic, but less well known than the prisoner's dilemma and the tragedy of the commons) is one on spatial distribution from the master himself, Schelling, in his book on *Micromotives and Macrobehavior* (1978/2006). The example does not have a snappy name, so I just call it "the case of the two dining rooms." How might women and men, say students at a college, distribute themselves between two dining rooms? In Schelling's words (p. 37):

> In the easiest case, all the men and all the women prefer a one to one ratio and will choose the dining room in which the numbers are most nearly equal. Suppose that there are 120 women and 100 men, that the women have to choose in advance, and that everybody knows that everybody prefers 50-50.

The optimal aggregate behavior should get us to 60 women and 50 men in each dining room. This is not what happens. All the men will end up in one room with 60 women; the other half of the women will dine in an exclusively female environment. Does this sound absurd? The micro-macro master explained plausibly, tracing the individual decision-making. We start with 60 women in one room, 60 in the other. Schelling (pp. 37–38) continued:

> Now the men arrive, and by the time three-quarters of them have arrived there may be 40 in one hall and 35 in the other. The later arrivals notice a slight discrepancy and choose the dining room with the more nearly equal number. In one room there are 60 women and 40 men, and in the other 60 and 35. The room with 40 men is slightly more attractive, and the next arrivals go there, and now there are 50 men in that room, 35 in the other. The difference is now more noticeable, and the next 10 men enter the hall with more men and there are 60 men and 60 women in that one, 35 men and 60 women in the other. The last 5 men much prefer the room with more men, and they make it 60 to 65 in that room, leaving it 60 to 35 in the other.

Schelling went on to describe the rational choices if men in the room with the disadvantageous ratio are free to change their minds. We can work it out for ourselves. The final score will indeed be 100 men to 60 women in one room, with 0 men to 60 women in the other. This is not a neutral result, not merely a curious case of rationally irrational behavior. Nobody of the 220 diners will be happy with the end result. The micro-macro analysis exposes a problem, a deviation from optimality. Schelling (1978/2006, p. 19) noted, "What makes this evaluation interesting and difficult is that the entire aggregate outcome is what has to be evaluated, not merely how each person does within the constraints of his environment."

In the case of the two dining rooms, it is easy to think of rational solutions to the problem of seating. However, these solutions require a shift in agency, from individual freedom of choice to a collective form of decision-making, introducing some kind of choice procedure designed to match the desired numbers at the macroscopic scale. We could draw lots, with equal numbers for room A and room B allocated to two pots— one for women to draw from and one for men. Alternatively, we could agree that the first 60 women and the first 50 men go to room A, while everybody else goes to room B. The solutions are mathematically trivial. Socially and emotionally, they may be tricky. As Hardin (1968) noted, we may need the recognition of necessity before we are ready to pursue the higher-order freedom of mutually agreed-upon mutual coercion.

Now, with our three examples at the back of the mind, we can turn to the case of the three R's in animal research.

3.1.2   Applying the Micro-Macro Analysis to Animal Research

3.1.2.1 The Individual Researcher's Perspective

At the microscopic level, we should consider the perspective of the individual researcher, the Principal Investigator. "Biomedical research in today's universities is usually carried out by groups consisting of a leader and 5-20 or so trainees," noted Hubel (2009, p. 161), one of our protagonists in the previous chapter. Indeed, this is how we organize the research: in small groups structured around the ancient master-apprentice relationship.

The prototypal trajectory of a biomedical researcher begins with an open-ended search for a field of interest, perhaps when she or he is a high-school student. Imagine this person is you. Even at this time, how-ever open-ended the intention, there are likely a number of factors that reduce your degrees of freedom in exploration (what society expects of boys versus girls, what your parents want, what your friends decide to do, etc.). For some reason, you get stuck on the topic of consciousness—how fascinating it is that there is something like you, a creature that can think and see the things you do. You want to know more about it. You read a couple of books and figure out that neuroscience is the field for you. You go to college and take a set of vaguely relevant courses; at least you are accumulating credits. Now you need to choose a laboratory. Which do you choose?

The laboratory you choose will have a huge impact on your trajectory. Maybe you try out a couple of labs; you check which of the labs seems the most fun (where are the cool people), active (as you can glean from the website), or just really, really interesting (probing consciousness with an electrode!). You had not thought about doing research with animals, but you also had not thought about *not* doing research with animals. You have no opinion about the ethics. You are genuinely interested in the science, so you start working in the lab, probing a monkey's prefron-tal neurons with an electrode while it is performing a memory-guided eye movement task. The people in the lab are very nice, and they help you; they cover up for you when you make mistakes; and they hang out with you, drink beer, and eat pizza with you.

The research itself is hard, with some nasty parts; you cannot deny that, but it pays off. You get a Ph.D. degree, and you work as a post-doctoral researcher in one of the best places in the world for this kind of research. You do not just learn the trade; you become one of the bosses. You publish a few papers in prestigious journals. You get offered an assistant professorship at a university not too far from your home-town. Of course you take it. You continue working, improving your trade with the eye-moving monkeys. You are now adding probing tech-niques that even your mentor did not master.

In the early days of your training as a researcher, you had to take a bioethics course as well as a more specific training on the care and use of animals at the research institute where you were then working. The first thing you heard was the three R's. You still know them by heart, and you agree with the concept. Of course you do. You are a scientist; you want only the best for science, and removing any unnecessary animal suffer-ing just makes good sense. By now you start getting irritated a bit with animal rights activists, who think they know better than you and do not listen, and who believe you are just in it for the money or even that you get some sadistic kick out of torturing animals. In fact, it is the animal rights activists who are torturing animals—*human* animals—with their terrorist tactics.

You just want to carry on with the work. You do what you do best. Now that you are the Principal Investigator in your lab, you write research proposals and occasionally you get funds. The ethics applications look very much like the research proposals you are submitting to various funding agencies. Things move forward. It is nonsensical to think you could do the same experiments and use the same techniques on any other species than the nonhuman primate. Anyway, that is not even the question with the three R's. The question is, "How can you study consciousness with non-sentient tissue?" The answer is, "Not." You honestly see no possibility of Replacement in your research. The only Replacement you can think of would be to stop doing your research altogether and move on to something completely different. That would not make sense. It would hurt the science. Your skills would be wasted. It would damage your productivity, and even put your career at risk (although you are not worried about that; you have enough confidence to change direction; your career really is not the problem).

Replacement being impossible, you do what you have to do with respect to Reduction and Refinement. You do it happily, wholeheartedly. It is not even a question of being forced by ethics. Your own free will dictates the same. You use the minimal number of animals to get significant results. Without significant results, you cannot publish anything, so of course you want significant results. Who does not want significant results? You make sure you get samples that give you significant results, without spending unnecessary time and effort. Why would you get more than you need? You have better things to do, other research to conduct. There is not the slightest possibility you will do anything other than perfectly apply Reduction.

The same is true for Refinement. You are not lying when you say that you are truly concerned about your animals, giving them the best possible care (even an enriched cage), and minimizing the suffering by improving your procedures. Perhaps you get slightly confused about the meaning of Refinement, thinking it also applies to the scientific aspects of refining your techniques, not so much the removal of animal suffering. In any case, you are refining this and that and the other thing, all the time. (When the animal is "sacrificed," you may even scatter the ashes underneath that special tree by the river, while offering your prayer of attention; does that count as Refinement? Of character, maybe?)

You get a perfect score. You implement the three R's as best you can.

Congratulations, you just became full professor. You rule, all right. Who knows, you may even win the Nobel Prize.

3.1.2.2 Failure of the Three R's at the Macroscopic Level

By the time the Nobel laureate had become an emeritus professor, he could only lament the changes in research culture, particularly the lost time at the bench for lead researchers. For Hubel (2009), the present microscopic level was not microscopic enough. He made a valid point. Researchers really should spend more time at the bench and less time on administrative matters. Turning back the clock to a 1950s mode of operation, though, is not the right answer.

The problem is the myopic view, the inherent boundaries of the individual perspective. The myopic view and micromanaging works perfectly when it comes to carrying out difficult tasks that require specialist skills, but with eyes set on what is nearby we see very little of what is on the horizon. That is simply a law of perspective—one that cannot be changed by any trick or amount of training. Within one single brain, neither you nor I will ever acquire world-class specialist expertise in more than a domain or two. (If you are that very rare true genius maybe you can reach three, but certainly not four; I have never heard of four.)

Applying their myopic view, individual researchers organize their research in ways that conflict with the three R's at the macroscopic level. The Principal Investigator naturally is unable to absorb the vast collection of specialist literatures. As a result, she or he may simply not know about a relevant alternative approach to research, which does not use any sentient tissue. Then, Replacement fails. The Principal Investigator may know about the relevant alternative approach, but not have the necessary skills to apply it. Again, Replacement fails. The Principal Investigator may know about the relevant alternative approach and be willing to acquire the necessary skills, but he or she may simply not have the budget to revamp the lab accordingly. Once more, Replacement is doomed to fail.

The same issues emerge with respect to Refinement. Please read the previous paragraph again, substituting Refinement for Replacement and substituting "which causes less suffering in the animals" for "which does not use any sentient tissue."

It is a major challenge for any individual researcher to reorient the research trajectory; it costs time and money and personal effort, and it comes at the additional price of a (temporary) decrease in scientific output. In any given lab, the dynamics of hysteresis prevent Replacement and Refinement. There is a natural bias against changing the research trajectory. The Principal Investigator may rightfully claim it is better to continue the research trajectory in order to maximally advance science.

The animal ethics committee of the Principal Investigator's research institute asks, "Can you conduct this research in any other way?" The Principal Investigator hears a bit of emphasis on *you* and on *this*, "Can *you* conduct *this* research in any other way?" For two reasons, the Principal Investigator must answer "No." Logically, *this* research can never be *that* research. Apples are apples. Oranges are oranges. *This* is never *that*. And "you," the Principal Investigator, can only do what you do best.

Yet, the microscopic conclusion is not necessarily true at the macroscopic level. The identity of *you* becomes fluid. It can be anyone. If you cannot do it, maybe someone else can. The research objectives and methods become fluid too. There are always many possible research projects. The question is which projects deserve the time, money, and effort. Viewing the entire research field, the barriers against Replacement and Refinement should disappear. If there is a relevant alternative approach, it is simply a matter of investing in that kind of research. In fact, *not* investing in that alternative approach would represent an opportunity cost. (In Chapter 5, I explore the implications of opportunity costs in thinking about how to organize animal research.) At the macroscopic level, we realize that we have limited resources (e.g., budget from funding agencies) to conduct research; resources invested in one type of research cannot be invested in another. We have to choose. At the microscopic level, the Principal Investigator may feel there is no choice but to carry on as before. He or she will scramble to get the funds to move on, in the one and only possible direction: forward. The lack of other options may be true for that person, but not for the entire field. We do have a choice, once we transcend the boundaries of individual perspective.

Not only Replacement and Refinement fail at the macroscopic level. Each lab steers its own course, based on myopic vision, without knowledge or consideration of the research plans in other labs but with an idiosyncratic sense of "what is going on" in the research field. Some topics are hot and look attractive; others, not. As a result, we may see hotspots of research and conspicuous gaps. The investigations done in a group of labs follow the hype; they overlap excessively, in conflict with the Reduction principle. More animals are used than necessary to establish the given bits of knowledge. Conversely, some labs find themselves extremely isolated, studying processes and using methods that fail to connect in any meaningful way to the rest of the research field—to the point that the findings fail to advance science. Again, this is against the Reduction principle, with animals sacrificed for useless bits of knowledge.

The question I am raising here is one of spatial distribution: How are the research projects distributed across the entire field? If the distribution occurs by microscopic vision, we may expect distortions analogous to the case with the two dining rooms. In fact, we can readily find evidence of such distortions. To illustrate, I conducted a little survey on December 15, 2017 via Pubmed, the search engine of the U.S. National Library of Medicine. I compared the number of papers using the search terms "rodent" versus "primate" in combination with different brain structures and neurotransmitters. (I performed the same searches also with the terms "rat" and "monkey," with similar results.) I carefully selected brain structures and neurotransmitters that should be equally relevant to both mammalian orders: the brain areas hippocampus and thalamus, and the neurotransmitters acetylcholine and serotonin.

The combination of hippocampus-rodent (97,750 studies) was far more prevalent than the combination hippocampus-primate (37,663 studies). Conversely, the combination thalamus-rodent (16,558 studies) was noticeably less prevalent than the combination thalamus-primate (25,394 studies). For hippocampus, rodents are the preferred animals; however, for thalamus, the preference switches to primates.

The combination acetylcholine-rodent (41,709 studies) clearly took precedence over the combination acetylcholine-primate (25,295 studies). However, the combination serotonin-rodent (55,347 studies) was less common than the combination serotonin-primate (64,803 studies). For acetylcholine, researchers like to work with rodents; however, for serotonin, they would rather use primates.

The distributions are reliably different, as we can surmise through a conventional test of significant difference. (I converted the numbers to percentages, with 2×4 conditions, comparing the rodent distribution versus the primate distribution by Pearson's chi-squared test; chi-squared $= 30.4578$; df $= 2$; $p < .001$.) There is no rhyme or reason to this spatial distribution, at least not from a rational, macroscopic perspective. The existing biases are due to idiosyncrasy at the microscopic level—idiosyncrasy that, in each individual case, may involve perfect application of the three R's from the perspective of the Principal Investigator.

3.1.2.3 Scale Enlargement for More with Less

When Russell and Burch (1959/1992) proposed their principles, they spent quite a bit of time arguing that researchers have the proper tools available to estimate the numbers of animals they need to obtain reliable findings. Statistics provides those tools. For Reduction to be in accord with good science, the sample size in research should be large enough for investigators to be able to draw firm conclusions about the presence or absence of effects. Reason allows us to work out the statistical power as a function of sample size. To be able to detect a signal among noise or a meaningful pattern among irrelevant variation, we can estimate how much data we need to collect.

The good scientist develops an empirical question through theory (based on a single theory, or pitching different theories against each other) and generates relevant hypotheses. These hypotheses come with an expectation of what counts as the presence or absence of an effect. Statistics, however, does not tell us directly what counts. It gives us an estimate of the likelihood of observed differences under a particular theoretical model. Whether we take the observed differences to be meaningful, given the likelihood estimation, depends on us. Decision-making comes into it from outside, not from the statistics itself. We use conventional criteria to determine the significance (e.g., an alpha level of 0.05, or a chance of 1 in 20 of making a type I error, rejecting the null hypothesis even though it is true). Yet, too easily we forget that *we* manipulate the margin of differences that may reach statistical significance. We do so not only by setting the criteria for significance (e.g., the alpha level), but also by the sample size.

With larger samples, smaller differences turn out to be "significant." This also implies that in practice we can turn the question around and work with a self-terminating sample size. We can keep collecting more data until we reach a significant effect or until we finally convince ourselves that we should accept the null hypothesis. This approach creates a bias such that we are more likely to make a type I error (rejecting a true null hypothesis) than a type II error (accepting a false null hypothesis).

Using the minimalist, self-terminating approach to data collection, we risk concluding prematurely that effects are significant (thereby increasing the risk of failures to replicate). Our publication culture also makes it much easier to publish significant differences than null results. Thus, in two ways, our approach to statistics is out of balance. In our own research, we hunt selectively for significant differences; we are more excited to learn about significant differences than null results when evaluating the works of our peers for publication.

To be sure, this is hardly a new complaint. Even in the days when Russell and Burch were writing, statisticians raised concerns about low statistical power and publication bias (Cohen 1962; Sterling 1959). The issues flare up again occasionally, recently in papers by Button and colleagues (2013a, b; Button and Munafò 2014; see also a critical response by Quinlan 2013; the topic received an enlightening, ironic treatment by Friston 2012). The problems can only be fixed by changing the research culture. In Chapter 5, I address how this can be done—how this is in fact compatible with initiatives toward mega-science and open science. For now, I merely point out that, as with the case of the two dining rooms, the mathematics of the solution is trivial. All we need to do is shift from the micro-scale to the macro-scale to get the numbers adding up more nicely for the research. We can remove the imbalance in our testing. We can get more with less—more powerful data with fewer animals.

By lining up the research efforts across labs, we avoid spurious overlap and increase statistical power to obtain findings that are at once more robust and more precise. To give an example, five different labs—governed by microscopic vision—each conduct an eye-movement study using two nonhuman primates with slightly different procedures, both with respect to the experimental paradigm (e.g., stimulus conditions, recording techniques) and general care (e.g., housing, feeding, surgery). Each of these labs reaches the same overall conclusion: "The frontal eye fields play a role in perceptual decision-making." Unfortunately, the net result is excessive duplication of a relatively vague statement.

The data across the labs cannot easily be pooled to one data set due to the different procedures. Moreover, the set of five studies is too small for a meaningful meta-analysis. We just have five studies with the same overall conclusion, and some unsystematic variations in recording and treatment. The first paper would go a top journal. The second paper still has a good chance to get accepted for a first-tier journal, as it can be considered a fine replication. The third, fourth, and fifth have to scramble for publication in lesser journals, as the authors desperately try to add a unique angle post hoc. We have no use for replications of replications ad infinitum. Judged at the macroscopic scale, this is a poor application of Reduction. If, in contrast, the five labs had worked together, using the same procedures, they would have accumulated a more powerful data set—even using only half the total number of nonhuman primates. Using five animals, the one study, organized with macroscopic vision, draws stronger and more precise conclusions (e.g., with respect to the trial-by-trial co-variation between neural and behavioral dynamics) than the five individualistic studies using ten animals.

The advantages of pooling resources for statistical power make perfect mathematical sense. We get more with less by thinking at the macroscopic scale. For true Reduction, the research programs have to be orchestrated cooperatively across groups of researchers, not by individual researchers alone. I note that this implies a departure from Russell and Burch's (1959/1992) *The Principles of Humane Experimental Technique*, where the Principal Investigator was consistently characterized as an autonomous entity, free to carry out the research without regard for the research field as a whole.

3.1.2.4 Public Support as a Limited Resource

Principal investigators, applying the three R's in their own labs, unwittingly operate in ways that sabotage the three R's at the macroscopic level. This observation matters. To quote Schelling (1978/2006, p. 19) one more time, "What makes this evaluation interesting and difficult is that the entire aggregate outcome is what has to be evaluated, not merely how each person does within the constraints of his environment"—the "entire aggregate outcome," which is what people see from the outside. When members of the general public look at the pros and cons of animal research and when people outside the scientific community assess the returns of the investments, they naturally look at the whole.

The breakdown of the three R's at the macroscopic scale is lamentable enough as it is. It fails to remove animal suffering, but it also (at least implicitly) insults the general public. People outside the scientific community naturally expect researchers to organize their research rationally. The faculty of reason, after all, is supposed to be something all scientists are able, and eager, to use. Members of the general public, evaluating the aggregate, may not look kindly on the failures to remove animal suffering. Disappointment leads to loss of support. This is where the tragedy of the commons becomes a relevant warning.

We have a limited resource in animal research—one that may be considered renewable, much like a forest or a sea full of fishes, but also one that must be maintained carefully. The critical constraint for animal research (this may sound cynical, which it is, especially for anyone who truly worries about animals for their own sake) has not so much to do with the numbers of animals. Our faculty of reason points out that we can always breed more animals. It is not obvious that we would ever run out of animals to experiment on (though we might lose a few species to extinction, particularly if they are not good at mating in captivity). The real constraint comes from public support. I suggest that public support can literally be seen as a limited resource, which determines the types and the amount of animal research that can be conducted, in terms of legislation as well as financing.

Here and there, I notice some level of awareness among scientists that public support is indeed the critical currency for the organization of animal research (e.g., Holder 2014; Bennett and Ringach 2016). Typically, the awareness is aroused reactively, belatedly, by a sudden realization that animal research is under attack. One representative editorial, in the June 2015 issue of *Nature Neuroscience*, responded to the news that the "legendary" Nikos Logothetis "has decided to cease his work with nonhuman primates and will instead exclusively study neural networks in rodents" (p. 787). Logothetis had suffered a "never-ending stream of abuse" by activists that "should be rightly viewed as terrorists." The core problem would be that "a mistaken few with a very effective propaganda machine are misleading the public" (p. 787). The solution was public outreach. Scientists, the editorial argued, should simply correct the misguided view.

I agree, to some extent. Public outreach is crucial. However, any defense should be based on the faculty of reason, not on a blind belief in the righteousness of the status quo. Too often, researchers take a seat in the back when it comes to thinking about ethics. An older editorial in *Nature Neuroscience*, from the February 1999 issue, admitted as much: "Standards of laboratory animal care are undoubtedly better now than they were in the past, and it would be foolish to deny that these improvements are in part a response to the valid concerns of animal rights advocates" (p. 100). I do prefer the tone of this 1999 editorial (which, incidentally, also responded to "terrorism" targeted at researchers).

To gain public support, we must engage in open discourse—not just by lecturing, but also by discussing, negotiating, re-evaluating, and setting our priorities together. There are perhaps a few things researchers could—or even should—do differently in their research. Personally, I would like to go one step further by not taking the back seat on animal ethics, but by going to the front row. Let us take leadership in animal ethics, based on what we learn about animals through our research. The good scientist cannot be happy in that position of moral inferiority, waiting for cues from the general public before improving animal care. Instead of being a slow-learning "asshole scientist" (Miki Lauwereyns, personal communication, December 20, 2017), I would like to find out—proactively—what is the right thing to do based on reason and knowledge from data.

3.2   The Principles of Reasonable Experimental Inquiry

Researchers, duly following Russell and Burch's (1959/1992) principles, do not manage to reduce animal suffering as much as they should or could. The micro-macro analysis puts the finger on the problem of organization. The original guidelines no longer function as driving forces to improve the use of animals in research. Additionally, in Chapter 2, I noted that the progress in science and the changing societal attitudes toward animals provide a fundamentally different background that exposes inherent conceptual weaknesses with the three R's. We have a vastly increased appreciation for the mental life of certain animals, and we generally recognize that it is inherently wrong to cause animal suffering. In conflict with this changed context, the original formulation of the three R's used a speciesist rationale, in which animal suffering was not regarded as an insult to the suffering animal but to "humanity." Moreover, scientific research received absolute, unquestioned priority; regardless of its value, it was always assumed to outweigh any animal suffering. Russell and Burch further complicated the problem by erroneously insisting on an overgeneralized notion that scientific research and the removal of animal suffering naturally go together.

For all these reasons, I conclude we can no longer accept the original formulation of the three R's. In rejecting the original formulation, I propose a revision of the principles, effectively aiming to extract some of the good that the three R's seemed to promise but failed to deliver. As a first point of revision, I suggest the domain of the principles, as indicated in the title of Russell and Burch's (1959/1992) classic, must be changed from "humane experimental technique" to "reasonable experimental inquiry." The principles that I envisage are about an internally coherent pursuit of knowledge. The issue is not only the method ("technique") but also the objective. I use the word *inquiry* to include both. The experimental inquiry is based on our faculty of reason: sound judgment, fair, sensible, informed, reckoning logically. With our faculty of reason, we look to advance science. For this, based on what we know and avoiding hypocrisy in the ethics, we must take speciesism out of the equation and come to an organization of animal research that respects all animals, including the human ones, for their own sake. We will need to consider in each case which research justifies the use of how many of which kind of animal. This huge departure from Russell and Burch means that some types of research may not be judged valuable enough.

3.2.1   On the Valuation of Research

Causing animal suffering is wrong. It is wrong for the animal in question in the first place. It can also cause secondary distress when it hurts the feelings of others who care about the creature that undergoes the suffering. Unless there is something substantial to gain from it, our moral position opposes the causing of animal suffering.

Which kinds of gain justify which types of animal suffering?

Ultimately, we have to question the objectives of the research. Not all research overrides the concerns about animal suffering. Reasonable experimental inquiry works from the premise that the research seeks to obtain important knowledge—important enough to outweigh the animal suffering. Instead of taking the importance for granted, we question it.

Whether we like it or not, we have to make a comparison, performing some kind of valuation with multiple factors. We have to consider the extent of direct and indirect distress as well as the prospective gains. In Chapter 2, I already pointed out how it is practically impossible to track the different factors in animal suffering, let alone to convert them to numerical units for a scalable computation. Further problems of intractability arise when we try to assess the prospective gains of research. Yet, for reasonable experimental inquiry, we do have to weigh the pros and cons and make clear decisions. Some heuristics are in order. Let me give a new set of ten thoughts, adding to the ten I gave in Sect. 2.2.2.

1. Fairness demands consistency in animal usage across different areas of society. This means we must develop a coherent view of what is acceptable (and what is not) in using animals for research, in the food industry, as pets, or for display or entertainment.

2. Direct and indirect animal distress varies as a function of species. In practice, the weighing of animal distress involves ranking species, from morally wrong to morally worse.

3. To develop reasonable criteria for the ranking, we should rely wherever possible on scientific data about direct and indirect animal distress.

4. Reasonable is judged in the eyes of the general public. When organizing the research, we would be wise to target a healthy maintenance of the renewable but limited resource of public support.

5. The general public looks at the big picture. Getting the big picture to look reasonable requires thinking seriously about the micro-macro dynamics.

6. Reasonable experimental inquiry is itself a finite commodity. We can perform only a certain amount of research, constrained by time, money, and effort.

7. The finite commodity means that prospective research projects play a zero-sum game against one another. Some will win the investment; others will lose. In this competition for investment, reasonableness integrates ethical and economical considerations.

8. The zero-sum game pitches apples against oranges.

9. Reasonableness dictates that we include ethical considerations in how we judge the competition for investment. Put differently, if two research projects promise equally valuable gains but Project A would cause more animal distress than Project B, then Project B deserves to win. ("Equally valuable" also includes cases when the values of different gains are conceptually impossible to rank.)

10. This is also true if Project B and Project A are even more different than apples and broccoli—say, one targeting leukemia with mice and the other investigating Parkinson disease with monkeys.

That makes twenty thoughts already. Allow me to explore the issues further, and collect more thoughts, before working on the convergence. A pragmatically feasible proposal shall be shaped from this, but not yet; that is for Chapter 5. In Chapter 4, I would like to consider several cases in detail. Here, still in Chapter 3, I am about to offer my revised R's. However, first I should say a few words about the slippery, yet unavoidable, idea of ranking animals.

3.2.2   The Preferred Animal Model

Sometimes we forget the most obvious fact. In all discussions about Replacement, we tend to forget that in most cases the research with non-human animals is itself already a replacement. The whole purpose of biomedical research is to improve *our* quality of life and *our* longevity. The species we are principally interested in is none other than *Homo sapiens.* Yet, even though we aim to learn more about humans, we apparently feel the need sometimes to take a detour, developing the biomedical research using other animals than humans.

Here I am not considering animal studies that directly aim to improve the situation for the animals under investigation. I do not see any urgent ethical issue with studies that benefit the very same individual animals that are being studied (e.g., taking a blood sample from monkey Haruki to check his health condition). The ethical issue is with conducting research on animals—not for the sake of those animals, but for the sake of knowledge.

When do we replace humans by other animals? *That* is the first question. I think we replace humans too often and too easily, despite the undeniable proposition that *Homo sapiens* is the ideal animal model for our biomedical research.

Humans are by far the preferable subjects, not only because they (i.e., we) can give us the data we ultimately seek, but also because humans can give informed consent. True voluntary participation, with a proper understanding of the situation, is simply not possible without language. Only humans can give informed consent. To be sure, it may be true that "monkeys must be calm and willing participants if the experiment is to succeed" (Editorial, *Nature Neuroscience*, 1999, p. 100), but this most certainly does not constitute true voluntary participation with informed consent; rather, it is a form of forced labor in captivity, in which prisoners learn to listen to their guards through the mechanisms of operant conditioning.

Some sophists might claim that, even if the purpose of research with nonhuman animals is to obtain general knowledge, the research could be regarded as something *for* the animals when the knowledge indirectly benefits later generations of the same species. Such sophists would perhaps see themselves giving vicarious informed consent, taking a role analogous to that of a guardian who makes decisions for a child. The argument might seem too far-fetched to mention here; however, I have heard it, more than once, so I thought it wise to take the time here to reject it explicitly. This sophistry uses the same logic that we easily recognize as horrific when it is applied to human subjects (cf. vivisections during World War II).

Only when we aim to help the very same animal can we reasonably give vicarious informed consent. Would *we* want to be helped in a particular way? If the answer is yes, we have an ethically defensible rationale to assume the animal may also want it. In all other cases, we have to start from the premise that any research with a nonhuman animal amounts to introducing a cost in the equation—something morally wrong that must be outweighed by other considerations if we are to go ahead with the research.

The core issue—the essence of what is morally wrong about using nonhuman animals—focuses on suffering and distress. Here, as a thought experiment, we might fantasize about animals that actually enjoy being subjected to research, much like "the pig that wants to be eaten" (to quote a book title by Baggini 2005/2006, offering 100 experiments for the armchair philosopher). In a similar vein, researchers sometimes casually suggest it is nonsensical to worry about Replacement if "animals are provided a good quality of life" (Olsson et al. 2012, p. 334) or if animals "have a much better life than they would in natural environments" (Tannenbaum and Bennett 2015, p. 127): no distress, no worries. Although this is a perfectly reasonable, almost Zen-like, deduction, I propose we should never lightly accept that the quality of life for a captive animal is good enough to quell any doubts about the ethics. Why would we use animals instead of humans in the first place if the life were so good, or even much better? The safe assumption in the real world is that the pig does not want to be eaten.

So I return to the basic question. When do we replace humans by other animals? In practice, we can distinguish two broad sets of situations. On the one hand, we turn to nonhuman animals if the research causes too much distress or if it creates risks of adverse effects—to the point that we cannot obtain informed consent from human subjects. (Sometimes, even if we can find human subjects willing to give informed consent, there may be legal constraints.) On the other hand, we use nonhuman animals when the data simply cannot be obtained with humans, regardless of the issue of informed consent, for technical (or species-specific) reasons.

In both sets of situations, we should acknowledge the basic brutality of animal research—the fact that we forcibly use animals for our own purposes. If the choice to use nonhuman animals pertains to risks and distress rather than technical issues, the proper ethical position is to backtrack and work with human subjects. If the research is important enough, we will find human volunteers for it. If the law does not allow us to work with human volunteers, we should endeavor to change the law.

Unfortunately, there really are cases in which it is not feasible to conduct the research with human subjects. In those cases, we should consider carefully which is the appropriate animal model. This implies performing some kind of comparison, in which we consider the direct and indirect distress caused by the animal research as well as the suitability of the animal model with respect to the prospective gains in knowledge. Will we be able to obtain the relevant knowledge with a given animal species? How much distress will be involved? To answer such questions, science must be our guide.

In the search for the appropriate animal model, I completely disagree with Russell and Burch's (1959/1992) notion that a given level of distress is worse for lower vertebrates than for higher vertebrates. In the section *Pain and Distress* (*PHET, Chapter 2*), they wrote:

> In general, the lower animal is the slave of its own moods. Its behavior is very largely automatic, and we know that we ourselves are most vulnerable when our behavior is most automatic. No[r] can a lower animal obtain the precious relief of *verbalizing* its distress. Far from despising lower animals (as it is convenient to call them) for these deficiencies, we should logically treat them with special consideration. [italics in the original; my correction (the original had 'Now' instead of 'Nor,' but that did not make sense to me).]

Such special consideration works in the wrong direction. Taken literally, it should imply that nonhuman animals always deserve more consideration than humans—a proposition that clearly does not match with the rest of Russell and Burch's reasoning. Moreover, I do not know, as "we know" in the above quote, "that we ourselves are most vulnerable when our behavior is most automatic." (Instead, I think unconscious processes are… unconscious, and therefore not felt as pain.)

The consideration should be applied consistently across species, starting with humans. I suggest there is a fair basis by which we can worry more about human suffering than about nonhuman primate suffering, and in turn more about nonhuman primate suffering than rodent suffering. This basis refers to the quality of sentient life, the richness and complexity of mental processing. I concur with Frey's (2011, p. 186) view:

> In my view, what matters is not life but quality of life. The value of a life is a function of its quality, its quality of its richness, and its richness of its capacities and scope for enrichment; it matters, then, what a creature's capacities for a rich life are. … Although animal life is typically not as rich and therefore not as valuable as human life, some animals have lives that are more valuable than other animals.

Here, Frey addressed only the direct aspects of sentient life, including suffering. The complexity of processing is a critical factor. Complex brains can process more information over longer periods of time, with higher degrees of sophistication. If anything, this increases the capacity for sadness, sorrow, suffering, and distress. Primary pain sensations may be similar in higher and lower vertebrates, based on similar neural circuitry for basic processes, but certain secondary, higher-order forms of suffering may be limited to humans and other higher vertebrates such as nonhuman primates. Data from neuroscience and behavioral science allow us to estimate the complexity of neural and sentient processing in various animals. With such information, we can approach the tricky comparisons of animals in a reasonable way—with fairness, internally coherent.

Ranking the capacities for direct distress, science informs us that humans are to be considered the prime victims of basic and higher-order forms of suffering, followed by other primates and certain mammals such as dolphins and whales, and then other mammals, and so on. Of course, this is not the only factor we should look at. We must also consider the indirect distress, as when a human child is more upset by the suffering of a pet mouse than by that of a wild mouse escaped from the sewer (even if there are blatant biases in this indirect distress; we may try to correct those biases, but until that correction takes place, the mere fact of indirect distress still counts as a cost).

Finally, when choosing the appropriate animal model, the considerations of direct and indirect distress weigh in opposition to the potential gains. It is certainly a difficult equation. However, the important note here is that we have good reasons to rank various animal species not only with respect to scientific usefulness, but also regarding the associated costs of direct and indirect distress.

3.2.3 New Definitions

I think I am ready now to offer new definitions of the three R's. As principles toward reasonable experimental inquiry, my three R's target an integration of ethics and science, without speciesism. Instead of autonomous decision-making by the Principal Investigator, I advocate active overview and cooperative data management (from collecting to sharing) by the relevant research community (my topic for Chapter 5).

More strongly than in the proposal by Russell and Burch (1959/1992), I emphasize that the R of Replacement is really the one that must come first. The all-important question is whether the animal research promises sufficient merit. Of course, all factors in the research proposals must be taken into account, including the species, the numbers of animals, and the procedures for experimentation and care.

3.2.3.1 Replacement
The research community judges the research objectives and methods. Human volunteers are the preferred subjects. A proposed research project can go ahead to replace human subjects with the appropriate model of nonhuman animal subjects if and only if (a) there are no human volunteers available and (b) the proposed project is judged important enough in comparison with alternative research projects. The choice for the appropriate animal model weighs, for all candidate species, the costs of direct and indirect suffering against the potential gains. The higher the costs, then the higher the gains must be to warrant the research.

3.2.3.2 Reduction
The research community reduces the number of nonhuman animals used in research by orchestrating cooperative data management (from collecting to sharing), effectively pooling resources. The number of animals used in each research project is determined based on solid hypothetico-deductive reasoning and the associated estimates of required statistical power.

3.2.3.3 Refinement
The research community ensures that researchers give the best possible care to the animals in all aspects before, during, and after experiments.

REFERENCES

Baggini, J. (2005/2006). *The Pig That Wants to Be Eaten. 100 Experiments for the Armchair Philosopher*. New York: A Plume Book.

Bennett, A. J., & Ringach, D. L. (2016). Animal research in neuroscience: A duty to engage. *Neuron, 92,* 653–657.

Button, K. S., & Munafò, M. R. (2014). Incentivising reproducible research. *Cortex, 51,* 107–108.

Button, K. S., Ioannidis, J. P. A., Mokrysz, C., Nosek, B. A., Flint, J., Robinson, E. S. J., et al. (2013a). Power failure: Why small sample size undermines the reliability of neuroscience. *Nature Reviews Neuroscience, 14,* 365–376.

Button, K. S., Ioannidis, J. P. A., Mokrysz, C., Nosek, B. A., Flint, J., Robinson, E. S. J., et al. (2013b). Confidence and precision increase with high statistical power. *Nature Reviews Neuroscience, 14,* 585–586.

Cohen, J. (1962). The statistical power of abnormal-social psychological research: A review. *Journal of Abnormal and Social Psychology, 65,* 145–153.

Editorial. (1999). Science and terrorism in Europe. *Nature Neuroscience, 2,* 99–100.

Editorial. (2015). Inhumane treatment of nonhuman primate researchers. *Nature Neuroscience, 18,* 787.

Fehr, E., & Fischbacher, U. (2003). The nature of human altruism. *Nature, 425,* 785–791.

Frey, R. G. (2011). Utilitarianism and animals. In T. L. Beauchamp & R. G. Frey (Eds.), *The Oxford Handbook of Animal Ethics* (pp. 172–197). New York: Oxford University Press.

Friston, K. (2012). Ten ironic rules for non-statistical reviewers. *NeuroImage, 61,* 1300–1310.

Hardin, G. (1968). The tragedy of the commons. *Science, 162,* 1244–1248.

Hauser, O. P., Rand, D. G., Peysakhovich, A., & Nowak, M. A. (2014). Cooperating with the future. *Nature, 511,* 220–223.

Holder, T. (2014). Standing up for science: The antivivisection movement and how to stand up to it. *EMBO Reports, 15*(6), 625–630.

Hubel, D. H. (2009). The way biomedical research is organized has dramatically changed over the past half-century: Are the changes for the better? *Neuron, 64,* 161–163.

Kuhn, S. (2017). Prisoner's dilemma. In E. N. Zalta (Ed.), *The Stanford Encyclopedia of Philosophy* (Spring 2017 Edition). Available at: https://plato.stanford.edu/archives/spr2017/entries/prisoner-dilemma/.

Newport, F., & Himelfarb, I. (2013, May 20). In U.S., record-high say gay, lesbian relations morally OK. *GALLUP News.* Available at: http://news.gallup.com/poll/162689/record-high-say-gay-lesbian-relations-morally.aspx.

Olsson, I. A. S., Franco, N. H., Weary, D. M., & Sandøe, P. (2012). The 3Rs principle: Mind the ethical gap! *ALTEX Proceedings, 1/12, Proceedings of WC8, 29*, 333–336.

Quinlan, P. T. (2013). Misuse of power: In defence of small-scale science. *Nature Reviews Neuroscience, 14*, 585.

Roelfsema, P. R., & Treue, S. (2014). Basic neuroscience research with nonhuman primates: A small but indispensable component of biomedical research. *Neuron, 82*, 1200–1204.

Rollin, B. E. (2017). The ethics of animal research: Theory and practice. In L. Kalof (Ed.), *The Oxford Handbook of Animal Studies* (pp. 345–363). New York: Oxford University Press.

Russell, W. M. S., & Burch, R. L. (1959/1992). *The Principles of Humane Experimental Technique*. Wheathampstead: Universities Federation for Animal Welfare. Available at: ALTWEB http://altweb.jhsph.edu/pubs/books/humane_exp/foreword.

Schelling, T. C. (1978/2006). *Micromotives and Macrobehavior. Fels Lectures on Public Policy Analysis*. New York: W. W. Norton.

Sterling, T. D. (1959). Publication decisions and their possible effects on inferences drawn from tests of significance—Or vice versa. *Journal of the American Statistical Association, 54*, 30–34.

Tannenbaum, J., & Bennett, B. J. (2015). Russell and Burch's 3Rs then and now: The need for clarity in definition and purpose. *Journal of the American Association for Laboratory Animal Science, 54*, 120–132.

Understanding Animal Research. (2017, September 6). *Numbers of animals.* Available at: http://www.understandinganimalresearch.org.uk/animals/numbers-animals/.

United States Department of Agriculture. (2017, June 27). *Annual report animal usage by fiscal year*. Animal Plant and Health Inspection Service. Available at: https://www.aphis.usda.gov/animal_welfare/downloads/reports/Annual-Report-Animal-Usage-by-FY2016.pdf.

CHAPTER 4

The Monkey Question

Abstract The use of monkeys in research represents a particular area of controversy. The chapter offers a critique of arguments typically offered in public debate. I note the weaknesses of retrospective thinking, the erroneous appeal to necessity, and the unfounded bias against working with human volunteers and rodents. The issues are illustrated via three scientific paradigms for which there exist valid alternatives: the development of a neural control system for robotic arm movement, the study of perceptual decision-making, and the study of the cognitive mechanisms underlying cocaine addiction and relapse. The three paradigms allow us to reflect on the opportunity cost neglect by individual researchers. Compromised by a conflict of interest, some researchers put forth misguided claims insisting on the irreplaceability of their preferred animal model.

Keywords Animal Ethics · Nonhuman primate · Replacement Opportunity cost · Conflict of interest

In 1838, Charles Darwin famously jotted down an aphorism in Notebook M about Locke and baboon, which I will quote here exactly in the way Cheney and Seyfarth (2007) quoted it, rather tongue in cheek, on page 1 of their *Baboon Metaphysics*:

© The Author(s) 2018
J. Lauwereyns, *Rethinking the Three R's in Animal Research*,
https://doi.org/10.1007/978-3-319-89300-6_4

Origin of man now proved.—Metaphysic must flourish.—He who understands baboon would do more towards metaphysics than Locke.

The thought shocked virtually all of Darwin's contemporaries and probably still shocks a good many of ours. In a review of the history and impact of the theory of evolution, Ruse (2003, pp. 103–104) wrote:

> Of all the issues raised by the *Origin*, the "monkey question" is the one that absorbed the attention of Darwin's fellow-Victorians and caused them great doubt. If evolution is true, then are we humans little more than overgrown primates? Are we descended from apes, rather than created in God's image? People on both sides of the Darwinian revolution worried incessantly about the status of humans. Sedgwick, a lifelong opponent of evolution, harped on the subject until his death. Lyell, although in a position to move to evolutionism well before Darwin, could never truly get God out of his science because he wanted to retain special status for humans.

The *Origin*, of course, refers to Darwin's landmark 1859 publication *On the Origin of Species by Means of Natural Selection, or the Preservation of Favoured Races in the Struggle for Life.*

Not everyone understood the preservation mechanisms to be morally neutral automatisms. Among the rhetorical distortions on favored races, I note the peculiar recurrence of a type of cynicism that turns "the monkey question" into a satire of progressive thought. The gist is always the same. Treating monkeys like humans would be too insane for words. Hilaire Belloc, a member of the Women's National Anti-Suffrage League and a vocal Catholic, wrote an essay in his collection *On Something* (1910) with the disingenuous title *The Monkey Question: An Appeal to Common Sense*, in which he pretends to offer a "pro-Simian" plea to put adult monkeys legally on the same footing as humans. My best guess is that Belloc wished to ridicule similar pleas for equal footing by others he considered "lower creatures," such as women and Jews.

Another example is even more explicit. On 31 March 1939, the notorious French collaborationist writer Robert Brasillach (later executed for treason) published a letter in the French fascist newspaper *Je suis partout* [I am everywhere] on *La question singe* [the monkey question]. In the letter, addressed to "*une provinciale*" [a woman from the provinces], he called for the organization of "*un antisiémitisme d'état et de raison*" [an anti-Simianism of state and reason], punning on the word *antisémitisme* [anti-Semitism] and thereby likening Jews to monkeys. The trick was to evade the 1938 Marchandeau Law against hate speech (see Sanos 2013, p. 324).

Perhaps it is fair to remark that we are still worrying incessantly about the status of humans. In our language, we continue to use animals as images and similes to demean and disparage. Neither you nor I would like to be called a jackass, bitch, worm, weasel, dog, pig, or rat; nor would we like to be accused of being mousy, horsey, fishy, mulish, or sheepish; and surely we will deny hogging, dogging, skunking, ducking, bugging, or parroting (Shepard 1978). Animals empower our hate speech, simultaneously insulting the addressee and the subject used in the address. Mason (2017, p. 139) invented the word *misothery* specifically for the latter component—the negative views and feelings that humans have about nonhuman animals. In analogy to the word *misogyny*, he combined two Greek words for it—one meaning "hatred" and the other "animal."

Naturally, the worrying about the status of humans versus other animals is most acute with respect to our nearest neighbors, speaking in terms of phylogeny. The monkey question comes to the fore as the most controversial in any consideration of the use of animals in research (e.g., writing in favor of the use of nonhuman primates: Capitanio and Emborg 2008; Phillips et al. 2014; Roelfsema and Treue 2014; Zhou 2014; Camus et al. 2015; writing an appeal against the use of nonhuman primates: Quigley 2007; Conlee and Rowan 2012; Bailey and Taylor 2016; and writing an interesting, less common, opinion requesting more diversity in the usage of animal models: Yartsev 2017). The similarity between humans and nonhuman primates amplifies the inherent conflict. The increased scientific benefits come with increased ethical costs—an inflammable equation. Typically, the debate on the usage of nonhuman primates reflects a tug of war. Authors want either all or none—a status quo (essentially a free license for scientists to use nonhuman primates) or a phasing out of all nonhuman primate research. The only thing beyond all doubt is that the monkey question deserves attention.

Researchers who regularly use nonhuman primates have become aware that their work is under scrutiny and does not appear to have full support by the general public. Phillips, leading a collective of fourteen authors, wrote an important review article on "Why Primate Models Matter" (2014), realizing (p. 818):

[W]e are at a critical crossroads. Unless NHP [nonhuman primate] research is given the philosophical, emotional, and financial support and infrastructure that is needed to sustain it and grow, we are in danger of losing irreplaceable unique models and thus, our ability to continue to explore and understand, and develop preventions and treatments for numerous conditions that inflict great suffering on humans.

The authors pleaded for more support, using arguments that are notably similar to those offered by Roelfsema and Treue (2014) in another opinion piece, already discussed in Chapter 3. In fact, the two articles offer a number of parallels. Both are seriously concerned about the future of research with nonhuman primates, alarmed by dwindling public support. Both essentially argue for an unquestioned status quo. Both urge for improved communication with the general public, assuming that the public merely needs convincing. Finally, both appeared in specialist journals that naturally have a stake in nonhuman primate research, preaching for the converted rather than addressing the general public that needs convincing. The article by Phillips and colleagues was published in *American Journal of Primatology*. Here, I am pulling these two articles to the foreground as good (recent, comprehensive) representatives of the typical position in favor of nonhuman primate research. Let us take a closer look at the arguments to assess their validity.

4.1   A Critique of Arguments

The articles by Phillips et al. (2014) and by Roelfsema and Treue (2014) both started with the proposition that biomedical research, particularly research involving nonhuman primates, "has played a vital role in many of the medical and scientific advances of the past century" (Phillips et al. 2014, p. 801). Biomedical research is identified as one of "two main fields of human endeavor that have propelled mankind forward" (Roelfsema and Treue 2014, p. 1200, using an interesting archaism, but hopefully not intending to imply that the human endeavor only propelled the male half of *Homo sapiens*). Both articles set out to illustrate this grand proposition, offering a dedicated program to support a received truth. We will look in vain for self-critique in these articles.

Although the authors offered long reference lists, the reasoning itself operated in a decidedly unscientific fashion, never attempting to question the received truth or to assess the actual extent of the contribution from research with nonhuman primates. Instead, we get a library of examples, most of which do not offer any specific link between the research and de facto medical progress. The basic logical structure appears to be circular, starting from the assumption—or even prejudice—that research with nonhuman primates is vital for progress, then offering a list of published papers, followed by the (overgeneralizing) conclusion that, therefore, research with nonhuman primates is vital for progress. Somewhere along this circular reasoning there slips in another received truth about the future:

> Here we review key areas in biomedicine where primate models have been, *and continue to be*, essential for advancing fundamental knowledge in biomedical and biological research. (Phillips et al. 2014, p. 802) [My italics]

> As documented above, basic neuroscience research with nonhuman primates has been *and continues to be* of paramount importance for past *and future* medical progress. (Roelfsema and Treue 2014, p. 1203) [My italics]

This promise for the future, which is taken for granted, appears to be another variant of the fallacious reasoning by Russell and Burch (1959/1992), as exposed in Chapter 2:

Major premise: *Research with nonhuman primates is vital for medical progress.*

Minor premise: *Future project X is a case of research with nonhuman primates.*

Therefore: *Future project X will lead to medical progress*

It really should be obvious that such reasoning is erroneous. The appropriate major premise is that *some* research with nonhuman primates may have been instrumental for *past* progress. Past success is not a guarantee for future success. The crucial ethical task is to question the prospective costs and benefits of each individual research project as compared to alternative research opportunities. This task requires looking ahead, beyond individual labs—not only taking into account past successes but also considering the changing context, including the changing attitudes with respect to animals and the improved research possibilities without nonhuman primates.

4.1.1   The Past Is Not Always a Good Model for the Future

A changing context can shed a whole new light on past behaviors, to the point that they no longer serve as a model for what is the right thing to do in the future. Such is the nature of practical ethics, situated in time and place, in the prevailing norms of society. Catcalling and wolf whistling are now increasingly understood to be unacceptable, constituting forms of sexual harassment. Smoking in public has become recognized, in just a generation or two, as a health hazard we should aim to avoid and contain (particularly with respect to passive smoking). Those who express crude racist views generally have at least some awareness of how out of place they are in contemporary society (even President Trump declared that he is "not a racist" after an uproar over his vulgar remarks on immigration; Kaplan et al. 2018).

Sometimes past successes were obtained in ways we can no longer accept as a model for future successes. Slavery offers a notorious example. In the present day, it is hard to conceive; however, two centuries ago, slavery had highly educated intellectuals among its defenders who offered (distorted, deliberately selective) utilitarian arguments in favor of the continuation of a practice that had been in the human repertoire since prehistoric ages. The case of Thomas Cooper provides an interesting example of a thinker opposed to the cruelty of the slave trade (see his *Letters on the Slave Trade*, 1787), but otherwise unashamed in favor of "humane" slavery as a necessary means to achieve certain agricultural benefits that he presumed unattainable with free labor (see Cooper 1835; Kilbride 1993). Cooper, in his time, using a conservative understanding of the roads to success, spoke of a necessity of slavery. With our 20/20 hindsight, we can safely conclude he was wrong. Agriculture did not collapse without slavery.

The reasons why Cooper was wrong are obvious to us now. His thought produced two fundamental errors. First, he underestimated the moral insult of slavery. Holding on to an outdated view, he was unable to adopt new, more reasonable insights about human rights or ethics in society. Second, he overestimated the unique and irreplaceable economic role of slavery. Looking back at past successes, he mistook slavery to be a necessary cause, whereas in actuality it was a sufficient cause. Successes could have been obtained differently.

I do not mean to imply that Phillips et al. (2014) and Roelfsema and Treue (2014) are the Coopers of our day. My purpose is not to offer an analogy to slavery, but to take the (in present eyes somewhat grotesque) case of slavery as an opportunity to learn something, analytically, about potential errors of thought in offering a utilitarian defense of a practice uniquely based on past history. Here, we can note the twin dangers of underestimating ethical costs and overestimating the role of nonhuman primate research. Both dangers are clear and present in the two articles.

4.1.2   Misleading Claims of Necessity

In the article by Phillips et al. (2014) as well as in that by Roelfsema and Treue (2014), nonhuman primate research is considered to be a nec-essary, rather than a sufficient, cause for progress, without a shred of evidence. All the authors did was to point, like Cooper did, to past suc-cesses. Moreover, the authors in both articles wavered about *what* the research would be necessary for. On the one hand, they kept empha-sizing the "need" for nonhuman primate research in order to obtain concrete medical objectives. The research would be "of paramount importance for past and future medical progress" (Roelfsema and Treue 2014, p. 1203), to "develop preventions and treatments for numerous conditions that inflict great suffering on humans" (Phillips et al. 2014, p. 818). On the other hand, the authors in both papers made a point of referring to a host of basic research on cognition and behavior, in which there was not the slightest pretense of medical benefits.

In the case of Roelfsema and Treue (2014), this conflation of basic and medical research was particularly salient, as they railed against "groups opposed to any animal research" who "refocused their broad assault onto just "basic" (as opposed to "applied") research" (p. 1200). For this reason, Roelfsema and Treue focused their article on basic neu-roscience research with monkeys. Yet, for any appeal to the public, they kept returning to the medical benefits. The studies, they argued, "have also impacted on our understanding of brain disease" (p. 1201). This observation inspired the following sweeping statement (still on the same page):

> It is therefore a flawed approach to define a divide between studies that address fundamental neuronal processes for cognition and those that apply this knowledge to understand brain disease and to develop new treatments.

Notably, the authors harped on the supposed absence of a clear divide between applied and basic research rather than on the question of what is necessary for medical progress. Yet, ultimately, it is the claim of necessity for medical progress with which researchers solicit public support for the research. Now, by Roelfsema and Treue's proposal, because we cannot define a divide between applied and basic research, we apparently should accept that the label "necessary for medical progress" applies to *all* nonhuman primate research, however remote from medical purposes. It is a move away from utilitarianism toward sophistry: a trick of categorization to improve the appeal of nonhuman primate research. To me, this sounds deceitful.

The real rationale for basic neuroscience research is to gain fundamental knowledge about brain and behavior. If the general public thinks this is not a good enough reason to sacrifice monkeys, then we should not try to argue our way around the issue by a trick of categorization. Then, we had better not sacrifice monkeys for basic neuroscience research. In fact, it should be fairly straightforward to distinguish research that promises direct versus indirect medical benefits. In the case of direct medical benefits, we can identify the types of treatment under investigation (e.g., the effects of a drug or a type of implant). In the case of indirect medical benefits, we cannot identify types of treatment; instead, the search will be open-ended. The principle is straightforward; the implementation must be a question of organization.

Personally, I think basic neuroscience research in general is very important (considering the whole research field, without isolating the portion with nonhuman primates); it is arguably an area of research that deserves more support from the general public—not only for its promises in terms of medical progress, but also in terms of its potential implications in all aspects of understanding the human mind and behavior. It may provide useful information for engineering, economics, law, and cultural studies, among other areas of society.

Thus, I think a unique focus on medical benefits is counterproductive—and in some cases, misleading. We should keep a clear eye on the true costs and benefits of research. In efforts toward public outreach, it also means we must place the research in its appropriate context: medical, if that is indeed the primary target, or otherwise. Basic research seems to me a label not to be shunned, but to be defended. At the same time, though, we should acknowledge that, when it comes to basic research, a vast field of possibilities opens up—most of which does not involve working with nonhuman primates. Here, the prospects with nonhuman primate research should be weighed against those with other types of research. It will take some convincing for nonhuman primate research to win that competition. Yet, that is the proper criterion.

In the meantime, we should not abuse the concept of necessity. General claims that nonhuman primate research is *necessary* for scientific or medical progress are simply wrong. The absoluteness, stated unconditionally, is plain nonsense. Of course, there will still be plenty of it scientific progress and medical progress with work solely based on humans, rodents, and other creatures, even if we ban *all* nonhuman primate research. No hospitals will fail to operate if we outlaw all monkey work tomorrow. Human lives that we can currently save will continue to be saved. We will even be able to save more lives than ever before thanks to medical progress achieved without touching a single monkey. The real question is what *more* we can do with nonhuman primate research that we cannot do without it. Does the surplus justify the basic moral wrong of harming monkeys? Is it important enough beyond all the other, morally more defensible things we could do? *That* should be the bar.

Unfortunately, Phillips et al. (2014) and Roelfsema and Treue (2014) never asked themselves what could have been, and what can be, achieved without monkeys. At best, they showed sufficient, but not necessary, causes for progress. In fact, for a number of the past successes, it is quite feasible to point out that the relevant findings could have been obtained in other ways, not relying on nonhuman primate research (see Sect. 4.2 on missed replacement opportunities). Time and time again, the authors limited themselves to listing a set of articles on topics for which they think the nonhuman primate research provided decisive insights. The authors did not show that these insights were inaccessible through other means. Worse, the decisiveness of these insights remains by and large speculative, without any empirical data to estimate the actual contribution of research findings toward progress (see Bailey and Taylor 2016, for an extensive discussion). This is particularly problematic when the argument ultimately rests, as it does in both articles, on the generation of medical benefits.

To put it bluntly, the authors have no proof that the purported decisive findings actually caused any medical benefits. The repeated claims that nonhuman primate studies are "essential," "necessary," "important," and "pivotal" in biomedical research reflect an opinion—a belief rather than a scientific fact. I do not dispute that nonhuman primate research has generated plenty of data. My point is that these scientific facts do not by themselves establish whether, or to what extent, they contributed to any medical benefits. Instead, the authors seem to work from the observations that (a) there have been medical benefits and (b) there have been research findings with monkeys to argue that *b* was necessary for *a*. Logic and science will tell us that such reasoning is wrong. It is possible that *b* had only a trivial or negligible role in establishing *a*, that *b* did in fact contribute to *a*, although a similar contribution could have been obtained in another way (i.e., *b* was part of a sufficient cause), or that *b* in fact had a critical contribution to *a*, which was impossible in any other way (i.e., *b* was a necessary cause). We need relevant data to be able to tell.

In short, Phillips et al. (2014) and Roelfsema and Treue (2014) made claims of necessity that are misleading. Given that they never showed any actual link between insights and medical benefits, we may even question whether the referenced research truly represented a sufficient cause for medical progress, let alone a necessary one.

4.1.3   An Erroneous Preference for Nonhuman Primates

In at least one way, it is easy to discredit the notion that nonhuman primate research would be necessary for medical progress. The preferred animal model for biomedical research should be the *human* primate. Scientifically speaking, we can safely assert that any progress with nonhuman primate research should likely be achievable also, or even more efficiently so, with research on human subjects. It is only when we start considering the ethics of research that we begin to realize that some types of research are so "invasive" or harmful that we might prefer to refrain from conducting the research with humans. This means that the "necessity" argument in favor of research with nonhuman primates, though on the surface a scientific argument, in actuality reflects a particular ethical compromise. Although, scientifically, the best and most valid data would be human, there appear to be other considerations that pull us away from our preferred animal model of *Homo sapiens*.

Reading the article by Roelfsema and Treue (2014), we need to scrutinize long and hard to find any trace of a realization that nonhuman primate research is in fact a suboptimal surrogate toward medical progress. There are only a few implicit indications, such as when it is noted that, "In exceptional cases, it is possible to record the activity of single neurons in the human brain, such as during neurosurgical interventions in patients with epilepsy" (p. 1202) or that, considering local or systemic application of drugs, "With some exceptions, these methods cannot be used in humans" (also p. 1202). The authors implied here that the research is technically feasible with humans and would be of more scientific interest than work with nonhuman primates, but that it is unthinkable for ethical reasons.

In the article by Phillips et al. (2014, p. 804), the issue of nonhuman primates as a suboptimal surrogate for humans is addressed explicitly:

> Some have argued that human rather than nonhuman primates are the more appropriate, and ethically preferable, subjects for biomedical research (e.g., Quigley 2007). The logic behind this is that despite the similarities between humans and NHPs [nonhuman primates], small but significant biological differences exist. Therefore, conclusive results cannot be obtained from NHP models, and so the ethically preferable choice would be to experiment on a limited number of humans. This position seems intuitive on some level. However, we should be reminded that it is essential for the protection of humans that prior research be conducted on animals. The Nuremberg Code, written as a result of the Nuremberg Trial at the end of World War II, defines a set of research ethics principles for human experimentation and states animal studies must precede research on humans.

Muireann Quigley (2007) did indeed write a lovely article for *Journal of Medical Ethics* questioning the appropriateness of nonhuman primates as a model for biomedical research and offering the human model as the preferable one. Curiously, the reference was absent from the impressive list compiled by Phillips et al. (2014, pp. 818–827; a total of maybe 500 articles) in perhaps a Freudian slip; let me make double sure to add the Quigley paper in my list. The logic behind Quigley's position is much less convoluted than suggested here. The logic is that, to learn about humans and to help humans, the best subject is human. Quigley's position is not just intuitive; it is based on logic and scientific thinking, and includes careful ethical considerations.

The core disagreement is in the ethical assessment. Interestingly, both parties harped on the similarities between humans and nonhuman primates, but in markedly divergent ways. For Quigley (2007), the similarities in biology and behavior should be integrated in a coherent scientific and ethical approach. In contrast, Phillips et al. (2014) considered the similarities only with respect to scientific questions, not with respect to ethics. For the ethics, they reverted to a speciesist stance, with a categorical divide between humans and all other animals. Tellingly, Phillips and colleagues based the ethical reasoning on an outdated document that was generated impromptu in an entirely different context to address entirely different issues.

The Nuremberg Code is a set of ten principles formulated in 1947 by judges during the so-called "Doctors' trial," when reviewing the cases of physicians and scientists accused of cruel and murderous human experimentation in Nazi-German concentration camps (Shuster 1998; Ghooi 2011). The principles sit buried in a 900-page volume, Volume II, "The Medical Case," to report in detail on the trials of war criminals (1949, pp. 181–182; specifically, in a section on "Permissible Medical Experiments"). Phillips et al. (2014, p. 804) quoted the third principle:

> The experiment should be so designed and based on the results of animal experimentation and knowledge of the natural history of the disease or other problem under study, that the anticipated results will justify the performance of the experiment.

The quote (Nuernberg Military Tribunals 1949, p. 182) was reproduced fairly accurately (apart from a missing article and an added comma). Yet, the usage of the quote in an argument in favor of nonhuman primate research seems misguided for several reasons. To be sure, the purpose of the Nuremberg Code was to address and define the moral wrong of *involuntary* human experimentation.

The physicians stood trial not because they failed to use animals before humans, but because they tortured and murdered humans. The first of the ten principles set the program clearly (Nuernberg Military Tribunals 1949, p. 181):

> The voluntary consent of the human subject is absolutely essential.

The contemporary proposal of human subjects as the preferable animal model for biomedical research, as put forth by Quigley (2007), of course refers solely to *voluntary* human experimentation. The Nazi context is entirely inappropriate here.

The key ethical issue is whether involuntary nonhuman primate research should necessarily precede voluntary human research. The Nuremberg

Code offers no useful guide on that issue. Literally, the third principle seems to require that animal experiments should *always* precede human experiments. That, of course, is an absurdity. We can, and do, go straight to human experimentation in much biomedical research, certainly also in cognitive neuroscience. If we cannot take the Nuremberg Code literally, then how should we understand it?

I think we should be reminded not of the Nuremberg Code, but of a more accepted, contemporary set of principles. As Ghooi (2011) noted in a thorough critique, "The Nuremberg Code has fallen by the wayside" (p. 72), being an ad hoc document, mostly plagiarism, that has not regularly been reviewed or updated. Instead, the World Medical Association (2013) has adopted the Declaration of Helsinki; this set of principles has been formulated and revised numerous times, each time aiming to catch up with the evolving nature of human ethics.

The Declaration of Helsinki on the ethical principles for medical research involving human subjects, in its 2013 incarnation, comprises 37 paragraphs. Only one of these mentions animals (World Medical Association 2013, p. 2192):

> 21. Medical research involving human subjects must conform to generally accepted scientific principles, be based on a thorough knowledge of the scientific literature, other relevant sources of information, and adequate laboratory and, as appropriate, animal experimentation. The welfare of animals used for research must be respected.

Clearly, this Declaration does not define animal experimentation as a necessary requirement before proceeding to human experimentation. The Declaration requires the usage of our faculty of reason, based on what we know. It does not say we always need to sacrifice the monkey first.

Some people do say we need to sacrifice the monkey first. They just say it in absolute terms, working from an unquestionable human supremacy. Zhou (2014), for instance, wrote a two-page article for *Trends in Genetics* with the following suggestion (p. 477):

> The ultimate goal of biomedical research, whether it involves NHPs [nonhuman primates] or not, is to benefit humans. Thus, in cases in which the necessity of using NHPs is urgent, such as the treatment of a new and acute infectious disease, human needs should be prioritized before the 3R principles.

In Zhou's world, the similarity between humans and nonhuman primates would be exploited in a perfectly one-directional way, an undiluted form of hypocrisy. The similarity would be seized upon for science, but completely ignored for ethics, without the slightest justification, even tossing aside the three R's of Russell and Burch (1959/1992). In a mind-blowingly bizarre turn, Zhou speculated that the monkeys would feel like heroes who gladly sacrifice themselves for the benefit of humans (pp. 476–477):

> It is difficult to imagine whether those NHPs that have been used experimentally would be proud of themselves if they knew how substantial the contributions are that they have made to human health. I would like to believe that the hero-like feeling is not human-specific, but one cannot believe that without also believing that pain is not human-specific ether.

Zhou concluded with a stirring final note—worthy of a happy ending in a Marvel movie, complete with a bombastic musical score (p. 477):

> For now, and in the future, though, we must recognize and treat NHPs just as we treat veterans from other wars, for they are surely soldiers in the battle against diseases.

Shame on *Trends in Genetics* for publishing such arrogant crock! Here, if Zhou wishes to insist on the imagery of soldiers in battle, he had better get the allegory correct. The monkeys would be no hero soldiers, but powerless victims of vivisection. Surely, such victims do not feel proud about any contributions they are forced to make, which will cost them their lives. The feeling of pride in inmates of concentration camps tends to derive from their acts of resistance against the masters.

Could there be a profound cultural difference at work? Zhou, from the Institute of Zoology in the Chinese Academy of Sciences, operates in a context where the absolute prioritization of human needs before the needs of nonhuman primates seems generally accepted by the powers-that-be. It has been noted recently that "China is positioning itself as a world leader in primate research" (Cyranoski 2016, p. 300). Some Western neuroscientists appear to turn a jealous eye, or a drooling mouth, to the possibilities in the world's second largest economy (p. 302):

> Bob Desimone was similarly impressed with the speed at which China moves. As a neuroscientist who heads the McGovern Institute for Brain

Research at the Massachusetts Institute of Technology in Cambridge, in January 2014, he had a 'meet and greet' with the mayor of Shenzhen. In March, the mayor donated a building on the Shenzhen Institute of Advanced Technology campus for a monkey-research facility, and the centre's soon-to-be director, Liping Zhang, promised that it would be ready by summer. Thinking that impossible, Desimone bet two bottles of China's prized mind-numbing liquor, *maotai*, that it wouldn't be done in time. He lost. The group raised most of the $10 million needed from city development grants, along with a small input from McGovern, and soon the first animals were being installed in the Brain Cognition and Brain Disorder Research Institute. "This place just makes things happen quickly," Desimone says.

Further on the same page, Desimone was quoted saying the place will be "an open technology base. Anyone who wants to work with monkeys can come." Yet, Desimone would do well to bet two bottles of *maotai* against it being a safe haven for long. According to Deborah Cao, "the immunity that China's primate researchers have had to animal-rights activism could start to erode… People are starting to use Chinese social-media sites to voice outrage at the abuse of animals" (still Cyranoski 2016, on p. 302; see also Cao 2015).

I predict that Zhou's hypocrite scientists-who-know-no-ethics will lose their Marvel battle in the end. The people will want real heroes, with their heart in the right place—human volunteers and scientists-with-integrity. My faculty of reason continues telling me that the human subject is the appropriate model for research about humans. Humans can give informed consent. No other animals can. If we cannot find human volunteers or if we find it ethically unacceptable to do the research with humans even if we were to find volunteers, then (and only then) might we consider using nonhuman animals—not necessarily primates. In this consideration, scientific and ethical reasoning must be integrated in a coherent way. Similarity between humans and nonhuman animals is a double-edged sword. The more similarity we have, the better for science, but also the more problematic for ethics. The ethical costs must be weighed fairly.

4.1.4   A Prejudice Against Replacement with Other Animals

In the articles by Phillips et al. (2014) and Roelfsema and Treue (2014), I noted a peculiar resistance against replacing nonhuman primates with other nonhuman animals. Indeed, it often appeared as though Refinement was really the only R under consideration. Tellingly, Phillips et al. (2014) added a single reference with a narrow focus to their nod in the direction of the three R's (p. 803):

> Researchers must specifically address the 3Rs before any research project with NHPs is approved, and guidance is available to assist researchers in implementing the 3Rs (Refinement, 2009).

The guidance referred to here advised only on refinements in husbandry, care, and common procedures in nonhuman primates. For cases when research projects look possible with other animals than nonhuman primates, the authors offered a variety of reasons to continue with monkeys after all. My favorite excuse—a delightful instance of black humor—must be this one (Phillips et al. 2014, p. 805):

> Pigs faithfully recapitulate human atherosclerosis but their large body size makes them difficult and expensive to handle and maintain.

What does this imply for nonhuman primates versus rodents?

A more dominant strategy, used by the authors of both papers, was to reemphasize the similarity between nonhuman primates and humans. If a theoretical question can be addressed empirically with both primates and rodents, the authors reasoned that primates should always be preferred because their anatomy is the most similar to humans. This argument comes dangerously close to ruling out the possibility of Replacement entirely from the outset, and thereby implicitly dismissing vast amounts of work that is in fact being conducted with rodents and other non-primates on the very same topics listed by the authors. Such dismissal would be highly contentious, at odds with the research community as a whole. The vast majority of biomedical researchers and neuroscientists clearly do invest in other animal models, accepting these as valid scientific approaches to studying the diseases and phenomena of brain and behavior listed in the two articles.

Lest the authors find themselves isolated on the fringes of the research community, they do have to accept the validity of other animal models. In an apparent move toward concession, Phillips et al. (2014) and Roelfsema and Treue (2014) showed a noncommittal interest in peaceful coexistence, claiming that the different approaches with primates and other animals are "complementary," without actually aiming to design an integrated approach. Instead, the argument in favor of the status quo implies that the research with nonhuman primates should proceed as before, as an independent effort that explores the unexplored, beginning de novo. The authors never considered a proper Replacement strategy whereby, for instance, research (out of ethical considerations) always would start with either humans or rodents. In a second phase, it would shift from humans to rodents, or vice versa, to confirm basic principles and to study detailed mechanisms. In all cases, work with monkeys would be left as a last resort—a contingent third phase only for specific, critical instances when the data from rodents (or other non-primate animal models) prove to be completely incompatible with findings in humans.

4.2   Examples of (Missed) Replacement Opportunities

Among the domains in cognitive neuroscience where research with nonhuman primates provided decisive insights, Roelfsema and Treue (2014) listed something like a table of contents for a textbook: perception and perceptual organization; storage of information in working memory; decision making; sensorimotor transformations; neuronal representations of reward, punishments, and reinforcement; motor control (including a readout for neuroprosthetic devices); and so on. Readers unfamiliar with this research should be duly impressed by the list. The voice of authority betrays no doubt; the tone of confidence inspires trust and acceptance.

I was also impressed by a list like this in 1994, when I bought the big hardcover *The Cognitive Neurosciences*, edited by Michael Gazzaniga. At the time, I was a beginning Ph.D. student in cognitive psychology, studying the mechanisms of object- versus space-based visual selective attention—a harmless piece of science, collecting behavioral data from willing human subjects. Leafing through the big book, I soon got bored with my cognitive psychology for all the things it failed to say about brain and mind. *The Cognitive Neurosciences*, among other encounters and readings, pointed me in the direction of monkey neurophysiology. That, apparently, was where I had to be for the decisive insights I desired. In monkey neurophysiology, the promise said, we could study the mechanisms of brain and mind to a level of precision that was unattainable in any other way, with any other species, including human and rodent. Learning about those mechanisms—what science could be more important or more interesting?

Any ethical qualms I might have had about using monkeys (I do not remember having had many of those at the beginning) were easily allayed. I considered using animals for science more acceptable than using them for food—and I did use them for food, just like most people in society. Nature made me an omnivore. If that was unethical, then how should we correct nature? Aha, *with science*. So if science was about improving nature, then using animals for science *had to be* more acceptable than using them for food. That was ethics! In any case, I did not think about it very deeply when I joined Okihide Hikosaka's monkey neurophysiology lab as a postdoc, at Juntendo University in Tokyo, in 1998.

The ethical qualms came and went, in waves, after joining the lab—after getting to know the ins and outs of the research, while I was taking care of the monkeys, operating on them, putting electrodes in them, getting them to perform behavioral tasks, reporting the data, and understanding the outcomes for science, for society, for me and my fellow researchers, and for our monkeys. I started seriously doubting the rationale offered. There were (and are) several factors driving this doubt.

First, I became more concerned about the ethical costs of using monkeys by witnessing firsthand how clever these creatures are, how rich their mental life is, and how very real their suffering was in the lab. My earlier idea about using animals for eating versus science went out the window—perhaps in general, but certainly for monkeys in particular. I could never consider eating a creature like Haruki. Maybe the same restriction should be true for using monkeys in science.

Second, I became more skeptical about the actual value of the research findings, even when published in the most prestigious journals. I began to realize how the impact of the research was routinely overstated, as when researchers studying basic mechanisms of perceptual decision-making made spurious claims about the necessity of their work for medical benefits. The claims were logically false, if not plainly confabulated.

Third, and perhaps most damagingly, I noticed that there were alternative types of research with humans and rodents that offered decisive insights about mechanisms of brain and mind, on the very same theoretical topics. The alternatives offered insights that were no less decisive than what was possible with nonhuman primates and often obtained in ways that really did not depend, or did not have to depend, on previous work with nonhuman primates.

Ultimately, it was this third factor, of the alternative opportunities, that made me decide to step away from monkey research. Such a step away from working with nonhuman primates is certainly possible more often than proponents are willing to admit, without any disastrous loss of knowledge. Indeed, there will literally never be any loss of knowledge even if we discontinue all monkey research this exact minute. We may agree to declare that all scientific knowledge gained with monkeys to this very moment will remain valid as a legitimate source for all eternity, given that it was obtained in accordance with the ethical principles of its context, of its time. We lose nothing that we already have.

My concern is with new knowledge and new research. My ethics looks to the future. Proponents of monkey research would have to point to putative, anticipated gains, which are sure to remain beyond our reach if we engage in different types of research. Those anticipated gains should outweigh the alternative prospects, while further warranting the extra ethical costs, before we can consider the use of nonhuman primates a reasonable option. A proper argument along these lines, in favor of research with nonhuman primates, is not easy to make. The papers by Phillips et al. (2014) and Roelfsema and Treue (2014) never actually approached this goal. To illustrate the point about alternative opportunities, let me mention three examples.

4.2.1   Real-Time Control of a Robot Arm

Among the decisive insights from nonhuman primate research, Roelfsema and Treue (2014, p. 1202) mentioned the following:

> A final example is provided by several promising approaches for prosthetic devices, which capitalize on our understanding of the NHP's nervous system. One of the aims is to develop prostheses so that paralyzed patients can control an artificial limb with their thoughts. Proof of principle has been demonstrated in monkeys that learned to control a prosthetic arm based on neuronal activity in cortical areas involved in motor control.

At the end of this quote, the authors added a single reference to an article from a research group led by Andrew Schwartz (Velliste et al. 2008). The topic of "neuroprosthetics" seems to be a favorite with proponents of research using nonhuman primates. I remember vividly how, in their defense of monkey work, my colleagues in Leuven

(Janssen et al. 2014) used the exact same image I used in my critique of monkey work (Lauwereyns 2014) at a symposium and debate organized by the rector of the University of Leuven on "Non-human Primates in Biomedical Research: Science versus Public Opinion?" The image all of us liked so much was Fig. 2 in an article from a research group led by John Donoghue (Hochberg et al. 2012). It showed a sequence of four photos of human Participant S3, a 58-year old woman with tetraplegia, successfully controlling a neural interface system with a robot arm to drink coffee from a bottle through a straw. In the last of the four photos, she beams the most delightful smile.

My colleagues in Leuven wished to point out that the amazing advance was made possible thanks to preliminary work with monkeys. In contrast, I noted that the contribution of nonhuman primate research in this advance represented a clear example of research for which we had, and have, alternative options that move to the same goal. Between rats and humans, we do not need monkeys to develop neuroprosthetics—certainly not anymore; in hindsight, there are very good reasons to think that the same progress could have been achieved entirely without monkeys. A landmark paper from a research group led by Miguel Nicolelis, published in the previous millennium (Chapin et al. 1999), was the first to demonstrate "real-time control of a robot arm using simultaneously recorded neurons in the motor cortex" (as stated in the title). The title did not mention it, but the work had been conducted with rats. It had built on the lab's own previous research with rats on the mechanisms of sensorimotor encoding through analyzing synchronous neural ensemble activity (Nicolelis et al. 1995).

At this stage, with neuroprosthetics already on the map, the history of neuroscience tells us that the groups of Nicolelis and Schwartz proceeded with monkey research. Schwartz had often worked with monkeys, so that was no surprise. Why did Nicolelis shift to monkeys? One year after the groundbreaking rat paper, his research group published a follow-up paper with monkeys (Wessberg et al. 2000) "to reproduce, in a robotic device, the kind of complex arm movements used by primates to reach objects in space" (p. 361). Similarly, Velliste and colleagues (2008) argued that their work improved on the example of simple one-dimensional control of a robot arm in rats by showing a more complex, more dexterous type of neuroprosthetic control in monkeys. Thus, rats were presumed to be incapable of multidimensional forelimb movement—a presumption that we now know to be unjustified. Bonazzi and colleagues (2013) established that rat forelimb motor cortex distinguishes five classes of limb movements (abduction, adduction, extension, retraction, elevation) and four classes of paw movements (opening, closure, opening/closure sequence, supination). The scientific argument for shifting from rats to monkeys does not hold up to scrutiny.

Interestingly, Participant S3, she of the delightful smile, had already been implanted with a microelectrode array in her motor cortex on 30 November 2005 (Hochberg et al. 2012; supplementary Methods section)—several years *before* the purported "proof of principle" mentioned in the quote I gave at the beginning of this section (see also Hochberg et al. 2006). Clearly, it was possible to find human volunteers early on, and for good reason: They were the ones who stood to benefit from the targeted progress. In reality, between the rat research and the voluntary human experimentation, the primate work played, at best, a complementary role—one that happened to contribute a certain, untold amount to the neuroprosthetics. The research leaders in this field—Nicolelis, Donoghue, and Schwartz—all worked with monkeys to some extent. They supported each other's work, as good scientists do. However, this is not to say that the monkey research played a necessary role in the progress. The research culture, not scientific necessity, created a bias toward working with nonhuman primates. There is no convincing argument here to claim in principle that the same progress could not possibly have been achieved with humans and rats alone. There is certainly also no convincing argument here about any further need for research with nonhuman primates.

4.2.2   Perceptual Decision-Making

In the spring of 2010, my mind raced through a euphoric high and despairing low in the space of a few minutes when consuming the 20 May issue of *Nature*. It had been seven years already since I quit working with monkeys. Now I thought people agreed with my decision, hence the high—but then it dawned on me, as many times before, the bosses did not concur, ergo the low.

The research I had conducted as a postdoc in Okihide Hikosaka's lab, on neural mechanisms of perceptual decision-making and reward-oriented response bias (Lauwereyns et al. 2001, 2002a, b), had made me a whole lot wiser than I was several years earlier. It laid the foundation for two monographs I wrote a good while later (Lauwereyns 2010, 2012), but it also left me conflicted with the realization that all was not well on the ethical front. To study the neural mechanisms I was interested in, it was definitely possible to work with humans and rats. There were alternative options that cut the ground from underneath the feet of anyone insisting on the necessity of monkeys in this line of research. For me, it was a deal breaker. I could not continue with monkey research. As a "proof of principle" with respect to alternatives, for reward-oriented response bias, I developed a rat version of the task I had used with monkeys (Lauwereyns and Wisnewski 2006) and collaborated with rat neurophysiologists to decode a spatial signal in the brain of rats; it predicted their next move while they were sitting and waiting for their chance to move on with the task and obtain a reward (Takahashi et al. 2009, 2014). The closer I looked, the more evidence I found. We did not need monkeys.

Now, in that spring of 2010, I saw my thoughts appear in print, formulated in the reflections of others. Alison Abbott, a journalist with a science background, published a News Feature under the title "The Rat Pack," with the following little summary (2010, p. 282):

> Studying primates is the only way to understand human cognition—or so neuroscientists thought. But there may be much to learn from rats and mice.

Abbott opened her News Feature with the prominent mention of a name: "Anne Churchland had little time for rats" (p. 282). Here was a case of someone who had first worked on decision-making in monkeys for many years: "I didn't think rats would have the right sorts of brains," Churchland said (as quoted on the same page). Then, she visited Cold Spring Harbor Laboratory in New York, which was an eye-opening experience. She witnessed rats processing fuzzy sensory information and conveying their decisions through head movements. It proved to be a career-changing moment (p. 283):

> Churchland is so taken with the experimental possibilities afforded by rodents that she is staking her career on them. This summer she is moving to Cold Spring Harbor, where she will establish her own lab to study rat decision-making. Looking back, she wonders why she doubted rats' cognitive abilities so much. "They also have to make decisions in the wild in order to survive, and obviously have to accumulate and sift evidence to do so."

Churchland's hindsight speculation is misleading. She asked, "Why did I ever doubt the cognitive abilities of rats?" The problem was that she did *not* doubt. In her years working with monkeys, she had assumed that rats did not have the right type of brain. It was a presupposition, not based on reasoning, but on thoughtless acceptance. The presupposition, leaving no room for doubt, followed from the state of the art, the tradition, and the way research was conducted in neuroscience.

In a way, the presupposition was shocking. Doubt, as we know at least since Descartes, should be the engine for scientific thought. How can we get rid of our unwarranted presuppositions and apply a good dose of reasonable doubt? Sometimes, it takes a while (or even 13 years), but on a blue Monday we may observe something that does not fit with our mental model of the world. Churchland saw it with her own eyes: Rats were using sensory data to make decisions. She saw it, and it did not fit with what she had taken for granted until then. It was only at that moment she began to doubt. The observation with the rats threw her. She knew she had to investigate it further.

So I shall admit, I consider this move by Anne Churchland to be an inspiring example of a healthy and truly scientific take on the world, not afraid of self-criticism, with the guts to change things around. It positively made me happy when I read about it in that *Nature* issue.

My happiness did not last long. The sad news is that cases like Churchland's are so rare. In Abbott's News Feature, we can already see the defense mechanisms of the neuroscience community at work in the shape of Michael Shadlen, who was Churchland's former mentor (p. 283):

> "It is good to develop rodent models and see what they are capable of," says Shadlen. "But it still isn't clear to me that rodents do any serious deliberating in decision-making."

Shadlen implied that Churchland was making a mistake. Yet, if he really doubted the abilities of rats, his scientific spirit should set him on the path of investigation. He could stop his experiments with monkeys for a while to design relevant tests with rats. Of course, he will not do so. (Instead, a little demonic voice in my id feels tempted to replace "rodents do" by "Shadlen does" in the quote above.) As for deliberative decision-making in rats, the idea is not all that strange and can be traced back to the 1930s (Redish 2016; see also Steiner and Redish 2014, and

Papale et al. 2016, for a truly innovative approach to studying various aspects of deliberative decision-making—not merely translating from monkeys, but taking the lead with rats).

My euphoria on that day in May 2010 was further squashed by the editorial that *Nature* put in place to accompany the News Feature, in an apparent bid to preempt any ethical reasoning. Under the combative title, "Still prime time for primates," the editorial stated (p. 267):

> Will activists try to exploit these developments to argue that there is no longer a scientific justification for using primates?
>
> Even the most enthusiastic proponents of rodent-cognition research share this worry. They have had to argue hard to convince some people within the scientific community that useful work can be done in rats — and some sceptics remain unconvinced. But the rodent researchers have never argued that rats could or should replace primates.

Yet, that argument is inescapable. It is even implied in the very same editorial, a few paragraphs later:

> This approach [with rodents]—combined with the low cost of rearing and keeping rodents and the wide availability of genetic tools for studying them—promises to help scientists to reach these basic cognitive components with unprecedented speed and rigour. Rodent research is also a less ethically sensitive issue than primate research, so the more information that can be wrung out of rats and mice the better.

Nature may not like to admit it, but this is precisely an argument to replace nonhuman primate research with work on rodents. However, in the conclusion (still on p. 267), the editorial insisted that nonhuman primate research should proceed as before:

> Rodent studies have the potential to deliver reliable data that can inform human, and primate, cognition research, and allow those experiments to become even more revealing. But, if the goal is to understand the human brain and mind, rodent and primate work will need to be continued in parallel for the foreseeable future. It's a no-brainer.

Unfortunately, I reckon it really was a statement written without the usage of brains. With a proper scientific spirit, we note that we can do more with rodents than we previously thought possible. The rational corollary of this statement is that we do not need monkeys as much as we used to think. For each research project, we should carefully consider which is the appropriate animal model. The answer should be rodents more often than before. Rodent and primate work do not need to be continued in parallel. Rodent and primate work need to be carefully integrated in a coherent large-scale project with, for ethical reasons, a preference for rodent work wherever possible.

I am not ruling out all nonhuman primate work in all areas of science a priori. I argue against its necessity—and in many areas, I even question its usefulness—but I do leave open the possibility that we can occasionally expect sufficient usefulness from nonhuman primate research to warrant engaging in it, above and beyond what we can anticipate with other research options. In each case, though, before accepting this possibility, I definitely want better arguments than the "no-brainer" from *Nature*.

In the meantime, Churchland's career move proves very successful indeed, with research that helps shape the great new wave of research with rodents on the neural mechanisms of perceptual decision-making (Raposo et al. 2012, 2014; Brunton et al. 2013; Hanks et al. 2015; Licata et al. 2017; Akrami et al. 2018).

More generally, with respect to the topic of perceptual decision-making, I can draw, word for word, the same conclusion as in the previous section on neuroprosthetics. The research culture, not scientific necessity, created a bias toward working with nonhuman primates. There is no convincing argument here to claim in principle that the same progress could not possibly have been achieved with humans and rats alone. There is certainly also no convincing argument here about any further need for research with nonhuman primates.

4.2.3   *Cognitive Mechanisms of Cocaine Addiction*

My third example of unwarranted research with nonhuman primates is on the cognitive mechanisms of cocaine addiction. Even with respect to purely pharmacological research on cocaine toxicity, the preferred animal model should be human. With respect to functional magnetic resonance imaging (fMRI) research on the cognitive mechanisms of cocaine addiction and relapse, however, the choice to work with nonhuman primates is very difficult to follow. Work with monkeys has little or no ecological validity here because it completely disregards the complexity of the phenomenon of cocaine addiction, with its salient social and cultural determinants. Anyone who really wants to tackle cocaine addiction had better concentrate on the social-cultural dimension alongside the biological-pharmacological. We need human data for this, not monkey data.

In some sense, my third example is a virtual one because there have been no published reports of such research to date. I mention it because a research proposal on this topic by Wim Vanduffel at the University of Leuven generated considerable controversy in the early summer of 2013, and again later in that year (it also prompted the rector of the University of Leuven to organize the symposium I mentioned in Sect. 4.2.1.). In June 2013, Kim Van de Perre wrote an article in *De Morgen* (one of Belgium's leading newspapers) to report on public protest by animal rights activists concerning the proposed fMRI work with monkeys. In his defense, Wim Vanduffel was quoted as follows:

> Surely it is not because other universities don't do it that the University of Leuven has to stop? ... Anyway, who says the monkeys suffer in this? There is no proof of that. Then you might as well condemn all research with animals as animal suffering. Whether it involves monkeys or fruitflies. Legally we're doing nothing wrong. (My translation from Dutch)

It is a staggering bit of misinformation. Monkeys definitely suffer from being implanted with the kinds of head posts made of dental cement, necessary for the research. I also know there is a big difference between monkeys and fruit flies—even a legal one, in the sense that fruit flies are not part of any Animal Welfare Act on planet Earth. A few paragraphs after the above quote, Vanduffel sounded even more irrational:

> Look, this kind of protest by a few individuals makes me furious. It is even a crime against humanity, because scientific research with animals is for the benefit of humans. (My translation from Dutch)

I hope Vanduffel's level of reasoning was more professional, and less kneejerk defensive, in the actual research proposals for national funding as well as for approval from the Animal Ethics Committee of the University of Leuven. Remy Amkreutz, writing in November 2013 in *De Morgen*, reported:

> Now it turns out the ethical committee of the university, in which ten academics judge whether research can proceed, did not give a positive

recommendation. Still, the research went ahead, in part because the FWO [*Fonds voor Wetenschappelijk Onderzoek*; the Flemish Fund for Scientific Research] allocated a subsidy of 341,500 euro. (My translation from Dutch)

The case highlights some real-world barriers to ethical reasoning, to which I must return in the next chapter. To rethink the three R's in animal research, we will have to find ways to deal with this kind of catch-22 situation, in which funding agencies assume that ethical committees provide a thorough check, while these same ethical committees defer to the positive recommendations from funding agencies. In Chapter 5 I discuss how to organize animal research in line with the rethinking of the three R's.

As for Vanduffel's fMRI work with monkeys on the cognitive mechanisms of cocaine addiction, I see no sign of it yet. A search on 27 January 2018 via Pubmed (up to date, almost a week after the reopening of the U.S. government), more than four years after the newspaper articles in *De Morgen*, produced exactly two items with the terms "Vanduffel cocaine": two articles from before the research proposal, presumably included as key references in the applications (Mandeville et al. 2011; Nelissen et al. 2012). These two articles were in decent journals, but hardly world news—no more than a drop in the ocean in the increasingly difficult fight against cocaine addiction. In my view, the two papers were a waste of monkey lives, as is their virtual follow-up.

References

Abbot, A. (2010). The rat pack. *Nature, 465,* 282–283.

Akrami, A., Kopec, C. D., Diamond, M. E., & Brody, C. D. (2018). Posterior parietal cortex represents sensory history and mediates its effects on behaviour. *Nature, 554,* 368–372.

Amkreutz, R. (2013, November 21). *KU Leuven start twee nieuwe dierproeven met apen* [The University of Leuven starts two new animal experiments with monkeys]. *De Morgen.* Available at: https://www.demorgen.be/binnenland/ku-leuven-start-twee-nieuwe-dierproeven-met-apen-bc9e66ce/.

Bailey, J., & Taylor, K. (2016). Non-human primates in neuroscience research: The case against its scientific necessity. *Alternatives to Laboratory Animals, 44,* 43–69.

Belloc, H. (1910). *On Something.* London: Methuen. Available at: http://www.gutenberg.org/cache/epub/7354/pg7354.txt.

Bonazzi, L., Viaro, R., Lodi, E., Canto, R., Bonifazzi, C., & Franchi, G. (2013). Complex movement topography and extrinsic space representation in the rat forelimb motor cortex as defined by long-duration intracortical microstimulation. *Journal of Neuroscience, 33*, 2097–2107.

Brunton, B. W., Botvinick, M. M., & Brody, C. D. (2013). Rats and humans can optimally accumulate evidence for decision-making. *Science, 340*, 95–98.

Camus, S., Ko, W. K. D., Pioli, E., & Bezard, E. (2015). Why bother using non-human primate models of cognitive disorders in translational research? *Neurobiology of Learning and Memory, 124*, 123–129.

Cao, D. (2015). *Animals in China: Law and Society*. New York: Palgrave Macmillan.

Capitanio, J. P., & Emborg, M. E. (2008). Contributions of non-human primates to neuroscience research. *Lancet, 371*, 1126–1135.

Chapin, J. K., Moxon, K. A., Markowitz, R. S., & Nicolelis, M. A. L. (1999). Real-time control of a robot arm using simultaneously recorded neurons in the motor cortex. *Nature Neuroscience, 2*, 664–670.

Cheney, D. L., & Seyfarth, R. M. (2007). *Baboon Metaphysics: The Evolution of a Social Mind*. Chicago: The University of Chicago Press.

Conlee, K. M., & Rowan, A. N. (2012). The case for phasing out experiments on primates. *Hastings Center Report, 42*(suppl. 1), 31–34.

Cooper, T. (1787). *Letters on the Slave Trade*. Manchester: C. Wheeler.

Cooper, T. (1835). Slavery. *Southern Literary Journal, 1*, 188–193.

Cyranoski, D. (2016). Monkey kingdom. *Nature, 532*, 300–302.

Darwin, C. (1859). *On the Origin of Species by Means of Natural Selection, or the Preservation of Favoured Races in the Struggle for Life*. London: John Murray. Available at: https://en.wikisource.org/wiki/On_the_Origin_of_Species_(1859).

Editorial. (2010). Still prime time for primates. *Nature, 465*, 267.

Gazzaniga, M. S. (Ed.). (1994). *The Cognitive Neurosciences*. Cambridge, MA: The MIT Press.

Ghooi, R. B. (2011). The Nuremberg code: A critique. *Perspectives in Clinical Research, 2*, 72–76.

Hanks, T. D., Kopec, C. D., Brunton, B. W., Duan, C. A., Erlich, J. C., & Brody, C. D. (2015). Distinct relationships of parietal and prefrontal cortices to evidence accumulation. *Nature, 520*, 220–223.

Hochberg, L. R., Bacher, D., Jarosiewicz, B., Masse, N. Y., Simeral, J. D., Vogel, J., et al. (2012). Reach and grasp by people with tetraplegia using a neurally controlled robotic arm. *Nature, 485*, 372–375.

Hochberg, L. R., Serruya, M. D., Friehs, G. M., Mukand, J. A., Saleh, M., Caplan, A. H., et al. (2006). Neuronal ensemble control of prosthetic devices by a human with tetraplegia. *Nature, 442*, 164–171.

Janssen, P., Vanduffel, W., Vogels, R., & D'Hooghe, T. (2014, December 3). Research projects involving primates at KU Leuven: What, how and why? *Invited Talk at the Symposium and Debate on Non-human Primates in Biomedical Research: Science Versus Public Opinion?* Belgium: University of Leuven.

Kaplan, T., Weiland, N., & Shear, M. D. (2018, January 14). Hopes dim for DACA deal as lawmakers battle over Trump's immigration remarks. *The New York Times.* Available at: https://www.nytimes.com/2018/01/14/us/politics/david-perdue-trump-shithole.html.

Kilbride, D. (1993). Slavery and utilitarianism: Thomas Cooper and the mind of the Old South. *The Journal of Southern History, 59,* 469–486.

Lauwereyns, J. (2010). *The Anatomy of Bias: How Neural Circuits Weigh the Options.* Cambridge, MA: The MIT Press.

Lauwereyns, J. (2012). *Brain and the Gaze: On the Active Boundaries of Vision.* Cambridge, MA: The MIT Press.

Lauwereyns, J. (2014, December 3). Diminishing returns, increasing costs: Time for a paradigm shift? *Invited Talk at the Symposium and Debate on Non-human Primates in Biomedical Research: Science Versus Public Opinion?* Belgium: University of Leuven.

Lauwereyns, J., & Wisnewski, R. G. (2006). A reaction-time paradigm to measure reward-oriented bias in rats. *Journal of Experimental Psychology: Animal Behavior Processes, 32,* 467–473.

Lauwereyns, J., Sakagami, M., Tsutsui, K., Kobayashi, S., Koizumi, M., & Hikosaka, O. (2001). Responses to task-irrelevant visual features by primate prefrontal neurons. *Journal of Neurophysiology, 86,* 2001–2010.

Lauwereyns, J., Takikawa, Y., Kawagoe, R., Kobayashi, S., Koizumi, M., Coe, B., et al. (2002a). Feature-based anticipation of cues that predict reward in monkey caudate nucleus. *Neuron, 33,* 463–473.

Lauwereyns, J., Watanabe, K., Coe, B., & Hikosaka, O. (2002b). A neural correlate of response bias in monkey caudate nucleus. *Nature, 418,* 413–417.

Licata, A. M., Kaufman, M. T., Raposo, D., Ryan, M. B., Sheppard, J. P., & Churchland, A. K. (2017). Posterior parietal cortex guides visual decisions in rats. *Journal of Neuroscience, 37,* 4954–4966.

Mandeville, J. B., Choi, J.-K., Jarraya, B., Rosen, B. R., Jenkins, B. G., & Vanduffel, W. (2011). fMRI of cocaine self-administration in macaques reveals functional inhibition of basal ganglia. *Neuropsychopharmacology, 36,* 1187–1198.

Mason, J. B. (2017). Misothery: Contempt for animals and nature, its origins, purposes, and repercussions. In L. Kalof (Ed.), *The Oxford Handbook of Animal Studies* (pp. 135–151). New York: Oxford University Press.

Nelissen, K., Jarraya, B., Arsenault, J. T., Rosen, B. R., Wald, L. L., Mandeville, J. B., et al. (2012). Neural correlates of the formation and retention of cocaine-induced stimulus-reward associations. *Biological Psychiatry, 72,* 422–428.

Nicolelis, M. A. L., Baccala, L. A., Lin, R. C. S., & Chapin, J. K. (1995). Sensorimotor encoding by synchronous neural ensemble activity at multiple levels of the somatosensory system. *Science, 268,* 1353–1358.

Nuernberg Military Tribunals. (1949). *Trials of War Criminals Before the Nuernberg Military Tribunals Under Control Council Law No. 10, Volume II.* Nuernberg: Superintendent of Documents, U.S. Government Printing Office.

Papale, A. E., Zielinski, M. C., Frank, L. M., Jadhav, S. P., & Redish, A. D. (2016). Interplay between hippocampal sharp-wave-ripple events and vicarious trial and error behaviors in decision making. *Neuron, 92,* 975–982.

Phillips, K. A., Bales, K. L., Capitanio, J. P., Conley, A., Czoty, P. W., 't Hart, B. A., et al. (2014). Why primate models matter. *American Journal of Primatology, 76,* 801–824.

Quigley, M. (2007). Non-human primates: The appropriate subjects of biomedical research? *Journal of Medical Ethics, 33,* 655–658.

Raposo, D., Kaufman, M. T., & Churchland, A. K. (2014). A category-free neural population supports evolving demands during decision-making. *Nature Neuroscience, 17,* 1784–1792.

Raposo, D., Sheppard, J. P., Schrater, P. R., & Churchland, A. K. (2012). Multisensory decision-making in rats and humans. *Journal of Neuroscience, 32,* 3726–3735.

Redish, A. D. (2016). Vicarious trial and error. *Nature Reviews Neuroscience, 17,* 147–159.

Roelfsema, P. R., & Treue, S. (2014). Basic neuroscience research with nonhuman primates: A small but indispensable component of biomedical research. *Neuron, 82,* 1200–1204.

Ruse, M. (2003). *Darwin and Design: Does Evolution Have a Purpose?* Cambridge, MA: Harvard University Press.

Russell, W. M. S., & Burch, R. L. (1959/1992). *The Principles of Humane Experimental Technique.* Wheathampstead: Universities Federation for Animal Welfare. Available at: ALTWEB http://altweb.jhsph.edu/pubs/books/humane_exp/foreword.

Sanos, S. (2013). *The Aesthetics of Hate: Far-Right Intellectuals, Antisemitism, and Gender in 1930s France.* Stanford: Stanford University Press.

Shepard, P. (1978). *Thinking Animals: Animals and the Development of Human Intelligence.* New York: Viking Press.

Shuster, E. (1998). The Nuremberg code: Hippocratic ethics and human rights. *Lancet, 351,* 974–977.

Steiner, A. P., & Redish, A. D. (2014). Behavioral and neurophysiological correlates of regret in rat decision-making on a neuroeconomic task. *Nature Neuroscience, 17,* 995–1002.

Takahashi, M., Lauwereyns, J., Sakurai, Y., & Tsukada, M. (2009). Behavioral state-dependent episodic representations in hippocampal CA1 neuronal activity during spatial alternation. *Cognitive Neurodynamics, 3,* 165–175.

Takahashi, M., Nishida, H., Redish, A. D., & Lauwereyns, J. (2014). Theta phase shift in spike timing and modulation of gamma oscillation: A dynamic code for spatial alternation during fixation in rat hippocampal area CA1. *Journal of Neurophysiology, 111,* 1601–1614.

Van de Perre, K. (2013, June 5). *Protest tegen Leuvens cocaïne-experiment op apen* [Protest against Leuven cocaine experiment on monkeys]. *De Morgen.* Available at: https://www.demorgen.be/binnenland/protest-tegen-leuvens-cocaine-experiment-op-apen-bf87e368/.

Velliste, M., Perel, S., Spalding, M. C., Whitford, A. S., & Schwartz, A. B. (2008). Cortical control of a prosthetic arm for self-feeding. *Nature, 453,* 1098–1101.

Wessberg, J., Stambaugh, C. R., Kralik, J. D., Beck, P. D., Laubach, M., Chapin, J. K., et al. (2000). Real-time prediction of hand trajectory by ensembles of cortical neurons in primates. *Nature, 408,* 361–365.

World Medical Association. (2013). Declaration of Helsinki: Ethical principles for medical research involving human subjects. *Journal of the American Medical Association, 310,* 2191–2194.

Yartsev, M. M. (2017). The emperor's new wardrobe: Rebalancing diversity of animal models in neuroscience research. *Science, 358,* 466–469.

Zhou, Q. (2014). Balancing the welfare: The use of non-human primates in research. *Trends in Genetics, 30,* 476–478.

CHAPTER 5

Toward Reasonable Experimental Inquiry

Abstract Here I construct an integrative view with clear and realistic proposals for the policymaking with respect to the use of animals in research. Replacement comes first, as an inherently ethical principle. Where Replacement is not possible, the use of animals should be carefully managed on the basis of collective decision-making at the level of research communities and funding agencies, not at the level of individual researchers. This proposal addresses opportunity costs via a macroscopic approach to managing research. Collective decision-making enables communities to ensure that we invest our time, money, and effort in the most optimal way. Detailed suggestions are made on how to organize the macroscopic approach, compatible with the extant research communities and thinking with the concepts of open science and big science.

Keywords Animal ethics · Opportunity cost · Collective decision-making · Open science · Big science

In a fair society, we recognize the pursuit of social justice as a basic human need (Corning 2011). Even nonhuman primates show types of behavior suggestive of inequity aversion (Brosnan and De Waal 2003), a phenomenon that could be seen as a precursor to the more expansive types of fairness and justice valued by humans. Fairness gives us a different outlook on life and society than a bleakly utilitarian perspective. When we think of the standards of living or the quality and length of our

© The Author(s) 2018
J. Lauwereyns, *Rethinking the Three R's in Animal Research*,
https://doi.org/10.1007/978-3-319-89300-6_5

lives, we are concerned not only with material benefits, but also with the more intangible aspects of doing the right thing. Without justice, goodness, love, friendship, beauty, or truth, some of us (the present author, in any case) may not wish to live, regardless of all engineering and biomedical research.

Sometimes, some of us prefer to sacrifice personal benefits for goals that transcend our individual lives. Martin Luther King, Jr., famously connected this idea to fitness for living in a speech in Detroit on 23 June 1963 (as quoted by NCC Staff on the *Constitution Daily* blog, 2013):

> There are some things so dear, some things so precious, some things so eternally true, that they are worth dying for. And I submit to you that if a man has not discovered something that he will die for, he isn't fit to live.

The ability to sacrifice things may be one of the features that salvage our species. Even without making the ultimate sacrifice of our lives, we can do many sublime things if we think beyond our personal benefits. That is exactly what the concern for fairness and justice is all about. The appeal of (nonviolent, rational) animal-rights activists rests on the belief that this moral concern should be extended beyond the human species (Singer 1981/2011). This is not a random, sentimental, or groundless belief. It is a matter of thinking with science and integrity, which applies the knowledge from science not only in science but also in ethics.

Sadly, in the current divisive climate, the positions of animal researchers and animal-rights activists are consistently misrepresented as being mutually exclusive. In stereotypical discourse, both parties tend to paint demonic pictures of the opposition—heartless sadists versus brainless terrorists. I urge all animal investors to reject this irrational polarization, both the irrational excesses on the part of certain animal-rights activists *and* the irrational biases that sometimes creep into the pursuit of science. It must be possible to combine a deep concern for animals with a radical commitment toward good and important biomedical research. This can be done by fully embracing the role of moral leadership, putting scientific knowledge to the best possible use toward organizing a fair society. Ultimately, that should be science's *raison d'être*: a public venture to the benefit of society, where benefit translates to well-being and well-being cannot exist without justice.

A fair society will be a knowledge-based society, one in which science plays a central role. Science gives us various scenarios of what is likely to happen in different circumstances. Scientists engineer new possibilities and predict the associated costs and benefits. Policymakers, as representative agents in that fair society, optimize the decision-making with the information and opportunities offered by science. Arguably, among all the scientific disciplines, neuroscience takes a special role in this process because it actually studies the cognitive and neural mechanisms of decision-making, along with all other aspects of sentient life. Indeed, several authors have written compelling arguments that underscore this central position for neuroscience in society (e.g., Gazzaniga 2005; Churchland 2011). Martha Farah (2010) even edited a book with contributions by dozens of colleagues, defining the position as a separable interdisciplinary field called "neuroethics." With the central position comes a huge responsibility, which neuroscientists are only beginning to grasp. The spotlight is on neuroscience to deliver on its promise in the moral domain, contributing relevant information to help define what is right and wrong, what is fair and reasonable, in the courtroom, in the code of law, and in all human doings.

Hypocrisy naturally undermines any endeavor in the moral domain. The project of neuroscience, and biomedical research more generally, should stand as a model of critical thinking that is hypothesis-driven and empirically oriented. Perhaps these abstract statements sound innocent enough, belonging to the broad consensus of what science stands for; however, in practice, the scientific community fails to live up to these standards. Critically, this failure is not due to any individual. At the individual level, I believe most scientists are genuinely engaged in producing the best possible work. However, considering the entire field, I perceive a lack of collective reasoning in the way we operate as a community. The whole is less than the sum of its parts. Instead of a conservative attitude toward animal ethics, we can (and I believe we should) adopt a proactive strategy to promote reasonable inquiry with better-organized and better-justified animal research. In line with my notes in the previous chapters on how to rethink the three R's in animal research, I now turn to the pragmatics, identifying particular challenges and possible solutions.

5.1   REPLACEMENT AS THE CRITICAL R

In first proposing the three R's of Replacement, Reduction, and Refinement, Russell and Burch (1959/1992) intended a specific order, with Replacement as the first principle, followed by Reduction and then

Refinement. However, that order has routinely been ignored, along with the underlying rationale that first we should try not to use animals at all; then, if we fail in our effort of Replacement, we should use the smallest possible number of animals; and finally, in actually using that smallest possible number, we should treat the animals in such a way as to minimize the suffering. Reading the original principles, the order makes sense; however, in practice, researchers often jumble it up. Roelfsema and Treue (2014, p. 1200), for instance, mentioned the three R principles of "Replace, Refine, and Reduce," whereas Balls (2010, pp. 19–20) seemed to prefer an alphabetical order, with the "concept of *Reduction*, *Refinement*, and *Replacement*" (italics in the original).

In fact, as soon as we think of "the three R's" as a trio (like the three musketeers, the three sisters, or the three kings), we naturally view them on an equal footing. Worse, once we lose the sense of order in the three R's, we can find ourselves in a bind about what to do when we can honor one or two of the three R's, but not all three at the same time. This has led some researchers to focus primarily on Refinement as the one R that seems easy enough to include in any project, while claiming to follow the concept of the three R's (cf. Phillips et al. 2014, p. 803, with their telling reference to only Refinement when stating that "guidance is available to assist researchers in implementing the 3Rs").

For clarity, there really is one R that has to come first, and that is Replacement. Replacement, I note, is the only R properly concerned with animal ethics. Reduction and Refinement are about how to conduct animal research, given that we deem the research unavoidable—that is, when we find ourselves forced to go down the ethically problematic path of "brute science" (to quote the apt concept by LaFollette and Shanks 1996), using our powers to exploit animals for the benefit of new knowledge. Reduction and Refinement are about numbers and procedures. In many ways, these numbers and procedures should simply follow from the premise that the goal is new knowledge. Then science rules, and ethics has very little to say. Once we commit to conducting animal research, we had better use the *optimal* numbers and procedures, even if the statistics require a large number of animals and even if the procedures must be painful. Generally, researchers will be very good at targeting that optimality in numbers and procedures—not for ethical reasons, but for purely scientific ones, further spurred on by economic pressures.

Thus, I propose to pull Replacement to the foreground as the one critical R in animal research. To be sure, the Replacement that I have in mind is no longer the original one formulated by Russell and Burch (1959/1992). Here, I repeat my definition of Replacement, as I offered it at the end of Chapter 3:

> The research community judges the research objectives and methods. Human volunteers are the preferred subjects. A proposed research project can go ahead to replace human subjects with the appropriate model of nonhuman animal subjects if and only if a) there are no human volunteers available, and b) the proposed project is judged important enough in comparison with alternative research projects. The choice for the appropriate animal model weighs, for all candidate species, the costs of direct and indirect suffering against the potential gains. The higher the costs, the higher the gains must be to warrant the research.

To be sure, this is a somewhat subversive definition of Replacement, emphasizing that all animal research is by definition a form of substitution when the real target is *Homo sapiens*. We should work with humans as much as possible, then think very carefully when and why we should step away from our preferred animal model to a nonhuman one. When thinking about when and why, we have to consider the research objectives and methods—primarily the anticipated gains—in the appropriate context. Researchers tend to discuss the anticipated gains of their projects in an illusory void, setting the potential benefits against a background of nothing. The typical reasoning goes approximately as follows:

Proposal: *If we do Experiment M with monkeys, we will obtain Knowledge X.*

Alternative: *No other animal model can give us Knowledge X.*

Therefore: *We gain by doing that experiment with monkeys.*

Anyone versed (even only a wee bit) in economics will immediately cry, "Foul!" The choice is not between X and nothing, but between X and myriad other things. We have to consider the opportunity costs.

5.1.1   Eyes Open to Opportunity Costs

He who does not command sufficient goods to consume without any restraint, must husband his resources. In a twofold sense he is obliged to be saving. First, he must not leave unused any part of the means at his disposal nor of their useful content. He must realize from them the largest degree of utility which can be obtained without harm. In the second place he is forced to accept the even more momentous task of choosing between alternative uses. His choice must be a use which satisfies a need of maximum intensity. Invariably the more important uses are to be selected, the less important ones passed over.

This quote by the Austrian economist Friedrich Von Wieser (1914/1927, pp. 44–45), translated by A. Ford Hinrichs, makes for basic good sense (if we are willing to ignore the archaic sexism)—the kind of good sense that hardly needs any reference. If forced to make a choice, we had better pick the best option. Thus, we have to take the alternatives into account. Von Wieser is often credited with coining the concept of "opportunity cost" (e.g., Encyclopaedia Brittanica, Wikipedia), although glimpses of similar ideas are surely to be found in the works of many (if not all) theorists of economics writing in the decades or even centuries before the said Friedrich. I do not need to trace the history of ideas here, so I will zip across time and space to this contemporary, straight-talking definition of "opportunity cost" lifted from the website of *The Economist*:

> The true cost of something is what you give up to get it. This includes not only the money spent in buying (or doing) the something, but also the economic benefits (UTILITY) that you did without because you bought (or did) that particular something and thus can no longer buy (or do) something else. For example, the opportunity cost of choosing to train as a lawyer is not merely the tuition fees, PRICE of books, and so on, but also the fact that you are no longer able to spend your time holding down a salaried job or developing your skills as a footballer. These lost opportunities may represent a significant loss of utility. Going for a walk may appear to cost nothing, until you consider the opportunity forgone to use that time earning money. Everything you do has an opportunity cost (see SHADOW PRICE). ECONOMICS is primarily about the efficient use of scarce resources, and the notion of opportunity cost plays a crucial part in ensuring that resources are indeed being used efficiently.

Even in the field of bioethics, the decision-making improves by considering opportunity costs (Kirby 1986). Returning to the reasoning for animal research, the appropriate consideration before engaging in any empirical study is to set the potential gains with one type of study against the alternative gains we might anticipate with another type of study:

Proposal:	*If we do Experiment M with monkeys, we will obtain Knowledge X.*
Alternative:	*If we do Experiment H with humans, we will obtain Knowledge Y.*
Another alternative:	*If we do Experiment R with rodents, we will obtain Knowledge Z.*
Question:	*Which Experiment represents the best option?*

Even if we cannot obtain the same bit of Knowledge X with any other animal model than the nonhuman primate, the potential gain of X should be considered not against zero, but against the alternative gains Y (obtainable with humans) and Z (obtainable with rodents). The potential gain of X does not have a surplus value of $X - 0$, but it has a marginal value of $X - $ (the next best option).

Crucially, when computing the various utilities, proper ethical thinking requires that we incorporate all costs, including the purely financial costs of running the experiments, as well as the more elusive costs associated with the moral wrong of using animals (let me call this simply the "animal cost"). For this, we must approach the organization of animal research differently from before.

A traditional, impoverished argument (in line with Russell and Burch 1959/1992) would emphasize the gain of knowledge against the purely financial cost, effectively disregarding the animal cost (or even denying that there is such a thing as an animal cost), as well as ignoring the opportunity costs (and their associated animal costs). Thus, in the traditional view, if potential knowledge X is considered to be of greater value than the associated financial cost, the conclusion would always be to go ahead with the proposed experiment.

However, balancing X against only the financial cost is not right. We have to compare X, Y, and Z, while incorporating their respective financial and animal costs. At first, this may seem like an impossible, speculative exercise. Assessing the values of completely different gains would be like comparing apples and oranges. Plus, how do we factor in the animal cost? However, in reality, we do compare between different categories of things, especially if we can purchase only a subset of what is on offer. The whole concept of money works from the notion that the values of a great variety of goods and services can be translated to a common currency. Given a million dollars, how do you put it to the best possible use?

In scientific research, we can consume only a limited amount. The majority of this research is funded publicly. Society invests a certain portion of its scarce resources to finance the work of researchers and research institutions. We, the taxpayers, invest in science. We do so by allocating (or having our representatives allocate) a fraction of our taxes to funding agencies that award research grants. That pool of public money for research is very finite indeed. It is arguably considerably smaller than it should be, which only emphasizes further the need to use that resource wisely—the better to advocate for more investment. Every monkey study that gets conducted implies a choice against one or more studies with humans or rats. To argue that such investment is justified, we should consider those alternatives fairly—even if the research objectives in the different proposals are completely divergent, even if it means comparing the proverbial apples and oranges.

With respect to the animal cost, we can apply a minimalist, but very lucid, rule of thumb. If, in all other aspects, two research proposals seem of equal value, then the one with the least animal cost should get priority. When in doubt, choose the one with the smaller ethical load. If, from the perspective of science and the prospect of new knowledge, a research proposal with monkeys does not appear any more or less valuable than another proposal with humans or rodents, then the monkey proposal should be rejected for having a heavier cost in the moral wrong of using brute force on a creature endowed with particularly strong cognitive and sensitive powers.

5.1.2   Real-World Barriers to Replacement

To be able to consider opportunity costs, researchers must of course recognize that they *have* multiple options in the first place. Optimal, reasonable, ethically justifiable decision-making involves a careful examination of the possibilities of Replacement. Unfortunately, in the real world, there are several barriers that make such decision-making exceedingly difficult. It is not even a problem peculiar to animal research. People end up failing to consider opportunity costs in many different contexts (e.g., Northcraft and Neale 1986; Legrenzi et al. 1993; Frederick et al. 2009; Spiller 2011; Zhang et al. 2017). Typically, the opportunity costs remain implicit; often, people are not aware that they are transforming decisions of "Which one?" into "Whether or not?"

The decision-making tends to get focused on a single option because it is familiar, salient, or easily available. Researchers who routinely use a particular animal model will naturally plan on continuing such research because all their expertise is geared toward using that model; they understand it most fully, their labs are set up specifically for that animal model, and their careers and reputations are built on it. In fact, it is their life work and their livelihood. From the perspective of an individual researcher—say, Professor P, with twenty years of experience using nonhuman primates—it is practically unfathomable to switch to another animal model. Singling P out as the maker of a moral mistake and asking him (the majority of such cases happens to be male) to switch to another animal model, without any financial or institutional support to do so, might come down to requesting poor P to commit professional seppuku.

Simply forcing people to switch would indeed be "inhumane treatment of nonhuman primate researchers," to speak with a title of that *Nature Neuroscience* editorial (already mentioned in Chapter 3) written in response to the news on 28 April 2015 that Nikos Logothetis had been bullied out of monkey work by animal rights activists. Interestingly, that editorial also noted in an aside that Logothetis "will instead exclusively study neural networks in rodents"—a remark that sounds curiously like an instance of Replacement (p. 787).

I think the Logothetis case is perfectly illustrative of the real-world barriers to Replacement. Apparently, Replacement *was* an option for him. Yet, he did not initiate it, for reasons that remain implicit. He only would resort to it under duress, after suffering treatment that (in this aspect I completely agree with the *Nature Neuroscience* editorial) can only be viewed as unreasonable and unacceptable.

Replacement is naturally hard to initiate by researchers who have a vested interest in a particular type of research. The scientific argumentation suffers from a limited perspective and is compounded by a deep conflict of interest. The researchers know their own field, but they may simply not be aware, or capable, of alternative types of research. In defending their research, they are defending their way of living, their core identity, and their honor. For that, like any besieged animal, they will fight as hard as they can. Any challenge on the ethical front will focus their scientific mind. They will narrowly apply their expertise to provide all possible reasons in favor of continuing to do what they have always done. In this, they will be formidable opponents to anyone who dares to challenge them. After all, they are the world experts at what they do. Yet, when Professor P says, "There is no other option," he actually means, "This is the best I can do." He is not the world expert of everything in the world. What he knows best is not necessarily the most desirable thing out there—maybe there are better things to be done, by others, or even by P himself, if given the appropriate support.

To be sure, I would never wish to bully Professor P out of his job through any abusive campaign. Rethinking the three R's should not mean putting certain people out of their jobs. However, it might mean rethinking the contents of jobs. As for P, he may not be able (or not be in a position) to see the best option. He is only human after all, just like you and me. If P is mistaken about the opportunity costs (as he might be, given his limited and conflicted perspective), then I think we have to devise a solution that is agreeable to everyone—or at least to as many people as possible, including P. It cannot be up to P alone to decide what he will do with public money. It also cannot be up to P alone to decide how and when he uses various species of animals, even if he is not relying on any public money in the process.

5.1.3   A Shift in Agency

The animal research under consideration here, when discussing the three R's, represents a collective effort in the public domain for society. The mere fact that we publicly debate the pros and cons of various types of animal research toward scientific and medical progress underscores that no individual can go it alone when planning and executing a specific empirical program. The animal studies of concern here take place in research institutes, where investigators keep animals specifically to be able to use them as subjects for experimentation. Phillips and colleagues (2014, p. 803), in their defense of experimentation with non-human primates, acknowledged that this enterprise implies the need for accountability:

> In all research institutions, investigators must convince a set of independent experts that the work they propose to do is justified and will be performed appropriately. These committees (Institutional Animal Care and Use Committees), which are mandated by Federal law, function separately from the research team and the funding agencies and have the right and obligation to restrict or stop any primate research that the committee considers unnecessary, inappropriately designed or inadequately justified given

the effects on the study subjects. These committees must include non-scientists and representatives of the community (lay members), to ensure that community standards for ethics are followed.

In theory, this legal framework sounds rational and fair. In practice, it does not quite work this way. Having served in three different committees of this kind (the Animal Ethics Committee of Victoria University of Wellington, the Schools' Animal Ethics Committee of the New Zealand Association of Science Educators, and the Animal Ethics Committee of the Faculty of Arts and Science at Kyushu University), I have noticed how the reviews invariably focus on the issues of Reduction and Refinement, with little or no consideration of Replacement. Effectively, the review procedures leave no room for an assessment of the opportunity costs. Instead, the Principal Investigator submits a research proposal with a rationale that sets the anticipated gains against zero. The question asked is "Whether or not?" Any Principal Investigator worthy of the title should easily be able to convince the committee members that, yes, something is better than nothing. Accordingly, in all my years as a committee member, I have never once seen a professional proposal rejected in the final decision, although I voiced plenty of doubts on a good number of projects. The rejection rate was zero percent, except in the case of the Schools' Animal Ethics Committee of the New Zealand Association of Science Educators.

Clearly, any review process that has a zero rejection rate is lacking teeth. It is extremely difficult for an Animal Ethics Committee to exercise its official mandate. In fact, it is not always clear to me that the committees are interested, or have the luxury of investing much time, in exercising their mandate; the typical sentiment closely matches that asserted by Sally Thompson-Iritani when she noted the goal "is getting scientists back to the bench doing their research, and animal care specialists getting back to their animals" (Cornwall 2017, p. 434; see also Chapter 2). The committee membership for an Animal Ethics Committee invariably has a majority of people who are favorably disposed toward animal research and balk at the red tape. The lay members, usually only one or two, are carefully handpicked to be none too adversarial. From a pessimistic or cynical angle, such committees may seem primarily to provide lip service.

Worse, even in the rare case when an Animal Ethics Committee does return a negative verdict, we may find its authority thwarted, as in the case of the proposed research on the cognitive mechanisms of cocaine addiction at the University of Leuven (discussed in Chapter 4, Sect. 4.2.3). In the given example, the host institute chose to follow the money. The Principal Investigator had already obtained a significant grant from the relevant national agency to carry out the proposed research. Thus, a problematic proposal can slip through the cracks of a system that promises multiple checks, but values economics more than ethics.

There is a relatively easy solution, which (perhaps somewhat surprisingly) may be in line with the wishes of the Association of American Medical Colleges and like-minded groups that are lobbying for less oversight (groups for which Sally Thompson-Iritani acted as a spokesperson; Cornwall 2017). We can reduce the red tape by restructuring the review procedures for ethics. The ethical thinking should be at its sharpest where the opportunity costs can be assessed. This is not at the level of the Animal Ethics Committee in a research institute. By the time the research proposals get there, it is already too late to hold them back. At that level, the best that can be done is to monitor the experimentation in terms of Reduction and Refinement—not asking "Whether?" but "How?"

The proper ethical thinking should occur at a stage before the review by the Animal Ethics Committee. To organize reasonable experimental inquiry, the appropriate place to insert the ethical thinking is in the allocation of resources for research. Whenever a decision needs to be made about allocating funds or space by a research institute or a funding agency, *that* is when the opportunity costs can be assessed. It is a simple and straightforward solution that represents a significant change in the way things are done, but it should be feasible, even with the existing institutes and funding agencies.

Ethics should be considered whenever we allocate a grant or provide the means to set up a laboratory. In all such cases, we can consider different proposals, as we already do. We can let a number of Principal Investigators put forward their best ideas, as we already do. There is no conflict of interest in trying to put forward our best proposals. We just argue our cases. However, the winning proposal should be convincing—not only scientifically, but also ethically; not only compared against zero, but also compared against the other alternatives. In practice, this means adding an ethical evaluation to the assessment of proposals for research, both at the level of funding agencies and research institutes. The ethical accountability then should be shared not only by the

Principal Investigator, but also by any funding agency or research institute that allocates its resources. In awarding grants or material means, the funding agencies and institutes should spell out their rationale for selecting the winning proposal, with reports that explicitly integrate the ethics and the opportunity costs.

My proposal, or my vision, comes down to a shift of agency that situates the ethical accountability not only with the Principal Investigator, but also (or even more so) with whomever allocates resources to the Principal Investigator. The Principal Investigator would not even have to argue against Replacement or opportunity costs. It would be the task of the relevant committees of research institutes and funding agencies to assess which is the best option, in light of scientific as well as ethical reasoning. It is their job to consider Replacement and the opportunity costs. Such committees are perfectly placed to do so macroscopically by comparing the different proposals.

This shift of approach solves the catch-22 situation in which the buck gets passed around between Animal Ethics Committees and funding agencies. In my vision, the Animal Ethics Committees would serve a subsidiary role, merely monitoring the good conduct of researchers in carrying out their work. This places less administrative burden on both the committees and the researchers.

The key is to shift to a form of collective decision-making on the allocation of resources *with* ethical accountability. At the same time, this would be a shift away from conflicts of interest and opportunity cost neglect toward reasonable experimental inquiry.

5.2   Practical Wisdom, Informed and Integrative

The shift of agency that I am advocating for essentially aims to "move things on a little," in the words of Henry Spira, one of the most effective animal rights activists of the twentieth century (as quoted by Singer 1998, p. 53—a quote that was in turn quoted by Hursthouse 2011, p. 137, and now by me). Practical wisdom should rule, working with all the information we can get, in a way that is neither radical nor preconceived, but integrative and flexible. As it happens, there are already some exciting new trends in science whose dynamics are perfectly suited for the right thing also in terms of the three R's in animal research. Each of us can move things on a little by actively supporting these new trends, toward open science and big science.

5.2.1   Open Science

The concept of open science is beginning to take hold thanks to the efforts of a group of vanguard scientists, who have become keenly aware of the hypercompetitive nature of contemporary science and its adverse effects on the publication and research culture (Martinson et al. 2005; John et al. 2012; Nosek et al. 2012, 2015; Chambers et al. 2014; Open Science Collaboration 2015; McKiernan et al. 2016; Edwards and Roy 2017). In this critique of contemporary science, researchers from diverse fields have noted that the reward structure in terms of publications, grants, and jobs causes "a disconnect between what is good for scientists and what is good for science" (Nosek et al. 2012, p. 616). Publication is the critical currency, which in its current form does not perfectly reflect truth. Too often and too easily, it reflects a carefully edited truth (taken out of context so that it offers a distorted image) or a premature truth (which crumbles under closer inspection). The current publication culture also promotes the search for trivial truths—the kind of minimal messages whose long-term goal is merely to get published. Smaldino and McElreath (2016) dubbed it the "natural selection of bad science" (in the title of their article). Such publications have little or even negative value for science, but they count as important rewards for the authors; these rewards are the stuff that careers are made of.

With publication as the critical currency, we find the hypothetico-deductive model of the scientific method compromised by a range of questionable research practices, as neatly laid out in Fig. 1 of the article by Chambers and colleagues (2014, p. 5). Being a lover of words, I can contemplate the message even better by reading only the caption to that figure:

> Lack of replication impedes the elimination of false discoveries and weakens the evidence base underpinning theory. Low statistical power increases the chances of missing true discoveries and reduces the likelihood that obtained positive effects are real. Exploiting researcher degrees of freedom (*p*-hacking) manifests in two general forms: collecting data until analyses return statistically significant effects, and selectively reporting analyses that reveal desirable outcomes. HARKing, or hypothesizing after results are known, involves generating a hypothesis from the data and then presenting it as a priori. Publication bias occurs when journals reject manuscripts on the basis that they report negative or undesirable findings. Finally, lack of data sharing prevents detailed meta-analysis and hinders the detection of data fabrication.

Some of the issues raised here mirror the ones I discussed in Chapter 3 in terms of a mismatch between micro-motives and macro-behavior. The problems of low statistical power, a lack of data sharing, and the publication bias toward positive effects are fundamental to contemporary science, not limited to animal research. However, in the case of animal research, the questionable research practices are particularly damaging because they not only produce suboptimal science, but they also increase the animal cost. In other areas of science, the questionable research practices are primarily methodological and economical issues of how to get more truth with less investment. Those are urgent enough concerns as they are. In the case of animal research, the issues have an extra moral dimension because they complicate the three R's. The concerns, then, must be even more urgent whenever we contemplate using nonhuman primates, rodents, or whichever species of sentient creature in the pursuit of knowledge.

Indeed, it should be a pursuit of *knowledge*, not reputation.

To address the questionable research practices, something has to change in the way we work. This must involve a shift away from governance by micro-motives toward rational behavior at the macroscopic level. Research and research-related jobs will always be limited; therefore, this is naturally a domain for zero-sum competition. We cannot simply try to reimagine it as a nonzero-sum competition in which everyone wins. The allocation of resources for research and research-related jobs will necessarily have winners and losers. However, it should be possible to change the rules of the competition so that at least science, ethics, and society do not lose. How can we structure the research and publication culture so that it provides incentives for good, honest, fair, and morally responsible science? How can we catch the questionable research practices in the act and sanction them out of existence?

Luckily, we are beginning to put our heads together on this problem. A large enough number of people have noticed it. Perhaps the problem has reached some "critical density," as Schelling suggested with respect to pollution, in the section poetically titled *The Dark Side of Plenty* in his 1971 essay "On the Ecology of Micromotives" (p. 76). We can listen to each other's proposals, support them, and work our way from the bottom up—a grassroots initiative toward better practices. Many of the proposals converge on the concept of open science. I gladly refer to the excellent arguments, including detailed suggestions and action points, particularly in the papers by Nosek and colleagues (2012, 2015), Open Science Collaboration (2015), Chambers and colleagues (2014), and McKiernan and colleagues (2016).

The basic idea is that openness increases accountability in all aspects of the research. In today's research culture, the only openness is in the research publications; however, "a scientific publication is not the scholarship itself, it is merely advertising of the scholarship" (Buckheit and Donoho 1995, p. 59; I first encountered the quote in Nosek et al. 2012, p. 623). Advertising, of course, is not the kind of openness that gives the accountability we need. Proper openness involves not only sharing the news, but also sharing everything that contributed to establishing the news. We can move several steps closer toward good, honest, fair, and morally responsible science with the following:

1. *Open data.* Sharing the data is an obvious way to improve the pursuit of knowledge, with more analyses, meta-analyses, re-analyses, and double-checks. It increases the risks for those who would fabricate data.

2. *Open methods and tools.* Open methods and tools increase the reproducibility and provide opportunities for coordinated data collection (and thus larger sample sizes, more statistical power, etc.). They also enable double-checks on the quality of the methods and tools. This gives correction and improvement.

3. *Open workflow.* Accountability is certainly also needed for the processes toward obtaining the data. Not only the data are shared, but also how the data are collected, from hypothesis to design, application of the experimental procedures, and recording of the data. This includes all processing steps from raw data to data for analyses. A major approach here is to work with registered studies, which makes it possible to coordinate research projects and thus avoid inefficient redundancy. Perhaps most importantly, it creates a platform for the sharing of negative findings.

4. *Open publication.* Even the openness of publication can be opened wider (items 1–3 of Box 1 in McKiernan et al. 2016), such as by posting free copies of previously published articles in a public repository, depositing preprints of all manuscripts in publicly accessible repositories, and publishing in open access venues.

5. *Outreach.* Although not usually included in the concept of open science, I think efforts toward outreach, beyond the scientific community, have an important place here as well. This increases accountability toward the entire society.

The latter point about outreach and public engagement is usually made separately, disconnected from the concern about good science and geared rather toward maintaining, or even regaining, public support for research (particularly controversial types, such as animal research; e.g., Holder 2014; Bennett and Ringach 2016). It may help to shout and wave flags for a few hours in the March for Science, as many hundreds of thousands did around the globe on Earth Day, 22 April 2017 (Appenzeller and The *Science* News Staff 2017). It may help even more to try tirelessly to communicate the contents of science, where it comes from, and how it works, in ways that convey its value proposition, wherever we can.

Research and research-related jobs will always be the treasured prizes in a zero-sum competition, but we enlarge the purse and get a bigger pool of resources (and thus *more* prizes and *more* winners) if we manage to convince a larger segment of society about the usefulness of science and stem the rising tide of irrationality in society, which distrusts science and is swayed by demagoguery. With more investment, the hypercompetitive nature of science gets treated—perhaps not cured, but at least attenuated. That would certainly also make it easier to reduce the appeal of questionable research practices and stay true to our ideals that have a bigger, long-term vision.

For the five of the points listed previously, it should be fairly straightforward what you can do. In writing this, I realize I can do better on most points myself. Realization is the first step; perhaps I should include it as point zero, the point of departure. If you are reading this, then you should now also be at least at that point zero. We can help others get there too simply by talking about the issues and by showing how we are now attempting to do some of these five points.

5.2.2   Big Science

From openness, we naturally move on to working together—or it can go the other way round, the desire to work together naturally makes us open up about all aspects of the research. Whereas the concept of open science puts the emphasis on transparency (and with it accountability), calls for big science have a complementary ambition, of expanding and improving the research possibilities through scale enlargement. *Homo sapiens* is certainly capable of devising immense projects that integrate the efforts of vast numbers of conspecifics and (what is the antonym?) hetero-specifics, willing or not.

In September 2017, I had the good fortune of being able to marvel at the still-standing product of one such project in Giza, Egypt: the Great Pyramid of Khufu or Cheops, some four millennia old, which is the oldest of the traditional Seven Wonders and the only one we can see with our own eyes today. *Homo sapiens* probably used many non-willing conspecifics back in the days the pyramid was being built. Does the awesomeness of that end justify its means? By our current understanding of human rights, we must reply with a resounding "No!"—however awesome the object (and it really *was* awesome). Yet, even with the unacceptable politics that enabled its construction, the Great Pyramid continues to bear testimony to the fact that humans can achieve truly amazing results by integrating their efforts for a single aim. The monument provides proof of principle.

More relevant for the present discourse, recent proof of principle of the ability to integrate our efforts for science is provided by two famous and positively mindboggling achievements. Everyone should know these examples, but just in case (or just to prove that I can refer to precedents), I hereby introduce Exhibit A, The Human Genome Project (International Human Genome Sequencing Consortium 2001). Three billion U.S. dollars were invested and twenty universities and research centers from three continents participated, with hundreds of researchers, to sequence the nucleotide base pairs of human DNA to chart the human genome. This is the type of science that has the kind of impact that society wants. It produces knowledge that really matters in every which way—theoretically, philosophically, practically, and medically.

For Exhibit B, let me point to the Large Hadron Collider (my preferred reference is a *Nature* News Feature from 25 March 2010, written by Zeeya Merali that focused on the social dynamics of the collaboration using a silly but charming pastiche for her title, "The Large Human Collider"). More than ten thousand people worked on the particle-physics project by the European Organization for Nuclear Research, with a total budget of nine billion U.S. dollars. The collider, the world's largest single machine, is used for such things as "accelerating two beams of protons to nearly the speed of light and then sending them in opposite directions around a 27-kilometre track" on a fearsome collision course (Merali 2010, p. 482). My intellect cannot cope with much of the science involved, but I enjoyed reading a *National Geographic* News article by Ker Than (2011) about "Densest Matter Created in Big-Bang Machine."

Molecular biologists and particle physicists can collaborate. Just a few weeks ago, I read an important article on polygenic overlap across major psychiatric disorders by Gandal and many dozens of colleagues, including several consortia (2018)—showing that psychiatrists can do it too. Neuroscientists and biomedical researchers who use animals should equally rise to the challenge. There are voices calling for it (Koch and Jones 2016; The International Brain Laboratory 2017, with as corre-spondent the same Anne Churchland of the exemplary move described in Chapter 4, Sect. 4.2.2). Even governments are asking for big neuro-science, with the United States BRAIN Initiative, the European Union's Human Brain Project, and the China Brain Project. Yet, these projects (seemingly initiated from the top down, by politicians) target infrastruc-ture rather than directly focusing on gaining fundamental new insights into brain function (Grillner 2014; see also Rose 2014; Greely et al. 2016, for discussions on the associated human ethical and neuro-ethical challenges).

It is early days still for big science, certainly in neuroscience and biomedical research. If the faculty of reason gets to rule, initiatives are bound to emerge with increasing frequency, confidence, and ambition. More to the point, with respect to rethinking the three R's, it should be obvious that big science is exactly how we can shift from microscopic perspectives to rational organization of research at the macroscopic scale. Thus, the topic takes on an added urgency in the case of animal research. With large-scale integrative efforts, we can properly address the three R's, achieving more with less—better and more ambitious science with fewer animals. This is in line with the principle of Reduction as well as Refinement through collective efforts that are mutually supportive of devising better procedures, with more researchers per animal and more data extraction per animal. Even Replacement would get a better chance as researchers, in a collaborative spirit, obtain a clearer view on the hori-zon of opportunities. Awareness is the first step toward being able to avoid neglecting the opportunity cost.

We can move several steps closer to good, honest, fair, and morally responsible science with the following:

1. *Joint research.* We go beyond the traditional single-laboratory mode of operation by effectively making the research efforts con-verge onto a shared goal. This means not just commenting on the design and the analyses of each other's work (for a free ride

as co-author), but actually organizing the data collection across different labs for a large-scale data set, going beyond anything a single lab can achieve.

2. *Integrated research.* Even if we do not manage to contribute to a large-scale data set across labs, it is often possible to connect translational efforts across different species in a carefully orchestrated way. For example, an experiment with humans in one lab should correspond as closely as possible to an experiment with rodents in another lab, with the express aim of combining the two experiments in the pursuit of knowledge (and, hypotactic, publications).

The present research and publication culture often has researchers from different labs contributing to each other's publications. In this sense, we already see plenty of collaboration in science. However, just exchanging co-author credits for supplying code or materials, or for commenting on each other's work, does not make big science. Truly integrated or joint research achieves scale enlargement. It goes far beyond what can logistically be achieved in a single lab, even if that lab has acquired all the tools, materials, and animals it can hold. Research of this expansive kind is, unfortunately, very rare in neuroscience and biomedical fields. That has to change for a proper implementation of the three R's.

5.2.3   Free Thought and Coordinated Action

If anyone had asked him, while writing mathematical and mathematical-logical treatises, or while working in the natural sciences, what goal he had in mind, he would have responded there is only one question really worth thinking about, and that is the one of the right life.

The description here is of Ulrich, the main character in Robert Musil's great modernist novel *Der Mann ohne Eigenschaften* (*The Man without Qualities*, 1937, p. 263; my translation from German). I am writing these words in a decidedly good mood, aided by the sound of Seal's 2008 song *The Right Life*. I like Musil's description. I like how Ulrich sees only one concern for science, a deeply ethical concern.

The right life is not necessarily the longest life or even the most gratifying in terms of primary and secondary rewards, such as sex, poetry, wine, money, or power. In the beginning and in the end, we must give due to justice, fairness, and freedom—the moral domain. There must be some balance overall. That would be the right life. With Ulrich, I would say that is why I am interested in science. I think scientists can contribute to the right life—not by taking a seat in the back when it comes to ethics, but by exploring the frontiers and applying what we learn, without hypocrisy or double standards, in an integrated and informed way. The right life really is the only question worth thinking about in science. That is the question at the heart of the present monograph, this effort of rethinking the three R's.

Now, nearing the close of this work, I would like to offer my vision of animal research and a set of action points toward that vision. In a way, my vision is simple. In one sentence, it is a shift from individual to collective work, with free thought and coordinated action. For a fuller description, let me use the following fat paragraph.

We would never want to use nonhuman animals when the target species is *Homo sapiens.* We would always prefer to work with human volunteers and their tissue. When it is not feasible to do so, we would collectively decide whether to use a particular species of nonhuman animals instead. This decision integrates scientific and ethical concerns. Here, open science and big science rule, allowing us to consider opportunity costs carefully. In this vision, everyone in the scientific community actively and openly shares responsibility for the research. If anyone uses a pair of monkeys, this really means *everyone* uses that pair of monkeys. We fully realize we are in it together, and we do not hide it. To make this happen, we would restructure the research and publication culture, getting rid of perverse incentives and rewarding good practices. Reputations would be based on relevant skills, the quality of ideas, and the design and analysis of research, not publication count. Research societies and consortia, managing the budgets made available by funding agencies, would organize research proposal competitions. The research societies and consortia carry out the winning proposals collectively. The data sets are made available to every scientist. Authorship would be limited to proposals and analyses; no individual names are attached to data sets. Research societies and consortia would organize new review procedures to recognize individual skills by awarding prizes and licenses for all aspects of research, from tissue treatment to surgery, from programming to analysis, and from writing to outreach. These awards would replace the role currently occupied by publication count.

How difficult would it be?

To change things, we can begin with these action points:

1. Do not use animals unless you honestly think you have no other choice.
2. If you see no other way than to use animals, work with rodents or animals further away from humans on the phylogenetic tree (below I will say simply "rodents," but I always mean "rodents or animals further away"). Nonhuman primates are the last resort.
3. Never start any line of research with nonhuman primates. You start either with humans or with rodents. From humans, you jump to rodents, or vice versa—always skipping nonhuman primates. You only go to nonhuman primates if you cannot understand any inconsistencies between humans and rodents, and if this lack of understanding must be resolved before you can make further progress.
4. Try to do as much as you can for open science (see recommendations in Sect. 5.2.1).
5. Try to do as much as you can for big science (see recommendations in Sect. 5.2.2).
6. Talk to your colleagues about all of this.
7. Add an ethical loop to all your activities in the scientific community.
8. Call out ethical issues when reviewing journal articles. Make journal editors and publishers aware that they are complicit when publishing research with substandard ethical practices.
9. Call out ethical issues when reviewing grant proposals. Make funding agencies and review committees aware that they are complicit when allocating resources to research proposals with substandard ethical practices.
10. Call out ethical issues when reviewing any proposal in any business of your research institute. Make your colleagues aware that they are complicit when turning a blind eye to substandard ethical practices.

This is my list.

5.3   The Sleep of Reason Produces Monsters

I think I have spoken my mind on the three R's. Will it matter? That, of course, is not for me to decide. What I do know, practically and minimally, is that my thoughts have a better chance of reaching other minds when I speak them than when I keep them to myself. What these spoken thoughts will do in those other minds is, physically and metaphorically, out of my hands.

"I have a dream," says the optimist. "I have a little nightmare," I might reply, in a darker mood. That mood would take the shape of the famous etching by Francisco Goya (1799/1969), *El sueño de la razón produce monstruos* ("The Sleep of Reason Produces Monsters"), number 43 in the set of 80, with the title *Los Caprichos* ("The Caprices" or "The Whims," looking anything but whimsical). The image shows a writer or an artist, possibly Goya himself, with his head buried in his arms, leaning on his desk, catching a nap. Behind him emerge the nocturnal. I count at least six owls, nine bats, one angry-looking boss cat (perhaps a lynx, judging by the pointy ears), and, oddly, one koala with wide-open eyes and a goofy grin (perhaps tripping). The writer or artist's legs are crossed at the ankles, making for a rather uncomfortable and strangely carefree posture. He cannot sit like that for very long without either feeling a sharp pain in his back or losing his balance and dropping to the floor.

Perhaps my reading of the etching betrays that I am an optimist after all. The writer or artist will wake up soon enough. He is not afraid, and the nap is structured to be short. The monsters are due to imagination without reason; but, Goya added, united with reason, imagination becomes the mother of the arts and the origin of marvels. If I am right and the sleep of reason cannot last long, then the good stuff will come sooner than later.

My mood is not so dark. I am listening to Seal's *The Right Life* after all, today, the twenty-third of February 2018, the final day for the linear writing phase of this book (tomorrow it goes into postproduction, with only defensive editing and correcting). I seem to have a seven-year itch in scientific writing; it was exactly seven years ago when I started the linear writing phase for my previous monograph. Perhaps the date is not entirely accidental. Perhaps I aimed for it a bit—exactly 197 years since the death of John Keats, and, infinitely more trivially, the twentieth anniversary of my Ph.D. defense on "the intentionality of visual selective attention."

The mere act of writing, for me, is always a matter of optimism.

So, let me be an optimist.

If my rethinking of the three R's is reasonable—and of course, I think it is reasonable because the whole point of my exercise was to let reason speak—then I trust it will meet with your approval. From there, we can move toward that coordinated action. I will do (and am doing) my bit. It begins with speaking. Now, you speak, too.

If my rethinking of the three R's is unreasonable, then I trust you will enlighten me by calmly and rationally pointing out my mistakes. (Of course, you will need to be awake to convince me of my imbecility. I am not afraid of any hysterical no-brainer monsters. My reason will make my mind immune to them.)

Could my rethinking of the three R's be moot? Is it not worth your time?

Even the whole idea of animal ethics seems to need a defense in a world that has so many other things on its mind. How can I argue that animal lives matter if (incredibly, sadly) it remains to be settled that black lives matter or that any unauthorized pussy grabbing is a crime? Also, how can I argue ethically when I am not an ethical man? (A question once raised by my ex-wife, possibly in allusion to the Jesus quote in John 8:7 of the Holy Bible, or a collection of texts considered by Christians to derive from divine inspiration: "Let any one of you who is without sin be the first to throw a stone at her.") The defense, I offer, must be that reason goes beyond the individual, beyond the sinner. Naturally, I can only begin with my own person, doing what I can, and feel I must, based on what I have learned, and what I know to be true. The moving is with my own person—yes, taking the "I" in "I think" out of the closet and into the world. This must be a form of radiating out from a core, not religiously but physically, in reality. The rays go in any direction, simultaneously. It is wrong to wait with doing the right thing, in any domain, at any time—no matter who you are, or what you did before.

Among the many areas where the right thing remains to be done, I can think of the use of animals, more generally, in society. It has always struck me as rather bizarre that the three R's were only talked about for research. How many pigs and cows could we save with a bit of Reduction in our meat eating? Not necessarily going all or nothing, but starting with something fairly manageable—say, a bit less? (How many human lives would be saved in the process, with healthier diets, more farmland, bigger crops, and more food for the world?) There are indeed many areas where the right thing remains to be done. Then, with the imperative of the Spike Lee joint: Do.

REFERENCES

Appenzeller, T. (2017). An unprecedented march for science. *Science, 356*, 356–357.

Balls, M. (2010). The principles of humane experimental technique: Timeless insights and unheeded warnings [Special Issue]. *ALTEX, 27*, 19–23.

Bennett, A. J., & Ringach, D. L. (2016). Animal research in neuroscience: A duty to engage. *Neuron, 92*, 653–657.

Brosnan, S. F., & De Waal, F. B. (2003). Monkeys reject unequal pay. *Nature, 425*, 297–299.

Buckheit, J. B., & Donoho, D. L. (1995). WaveLab and reproducible research. In A. Antoniades & G. Oppenheim (Eds.), *Lecture Notes in Statistics 103. Wavelets and Statistics* (pp. 55–81). New York: Springer.

Chambers, C. D., Feredoes, E., Muthukumaraswamy, S. D., & Etchells, P. J. (2014). Instead of "playing the game" it is time to change the rules: Registered reports at *AIMS Neuroscience* and beyond. *AIMS Neuroscience, 1*, 4–17.

Churchland, P. S. (2011). *Braintrust: What Neuroscience Tells Us About Morality*. Princeton, NJ: Princeton University Press.

Corning, P. (2011). *The Fair Society: The Science of Human Nature and the Pursuit of Social Justice*. Chicago: The University of Chicago Press.

Cornwall, W. (2017). Revamp animal research rules, report urges. *Science, 358*, 434.

Editorial. (2015). Inhumane treatment of nonhuman primate researchers. *Nature Neuroscience, 18*, 787.

Editorial. (2018). Economics A–Z terms beginning with O. *The Economist*. Available at: http://www.economist.com/economics-a-to-z/o#node-21529616. Accessed 13 Feb 2018.

Edwards, M. A., & Roy, S. (2017). Academic research in the 21st century: Maintaining scientific integrity in a climate of perverse incentives and hyper-competition. *Environmental Engineering Science, 34*, 51–61.

Farah, M. J. (Ed.). (2010). *Neuroethics: An Introduction with Readings*. Cambridge: The MIT Press.

Frederick, S., Novemsky, N., Wang, J., Dhar, R., & Nowlis, S. (2009). Opportunity cost neglect. *Journal of Consumer Research, 36*, 553–561.

Gandal, M. J., Haney, J. R., Parikshak, N. N., Leppa, V., Ramaswami, G., Hartl, C., et al. (2018). Shared molecular neuropathology across major psychiatric disorders parallels polygenic overlap. *Science, 359*, 693–697.

Gazzaniga, M. S. (2005). *The Ethical Brain: The Science of Our Moral Dilemmas*. New York: Dana Press.

Goya, F. (1799/1969). *Los Caprichos*. New York: Dover Publications.

Greely, H. T., Ramos, K. M., & Grady, C. (2016). Neuroethics in the age of brain projects. *Neuron, 92,* 637–641.

Grillner, S. (2014). Megascience efforts and the brain. *Neuron, 82,* 1209–1211.

Holder, T. (2014). Standing up for science: The antivivisection movement and how to stand up to it. *EMBO Reports, 15*(6), 625–630.

Holy Bible. (2011). *New International Version.* Colorado Springs: Biblica. Available at: https://www.biblegateway.com/versions/New-International-Version-NIV-Bible.

Hursthouse, R. (2011). Virtue ethics and the treatment of animals. In T. L. Beauchamp & R. G. Frey (Eds.), *The Oxford Handbook of Animal Ethics* (pp. 119–143). New York: Oxford University Press.

International Human Genome Sequencing Consortium. (2001). Initial sequencing and analysis of the human genome. *Nature, 409,* 860–921.

John, L. K., Loewenstein, G., & Prelec, D. (2012). Measuring the prevalence of questionable research practices with incentives for truth telling. *Psychological Science, 23,* 524–532.

Kirby, M. D. (1986). Bioethical decisions and opportunity costs. *Journal of Contemporary Health Law and Policy, 2,* 7–21.

Koch, C., & Jones, A. (2016). Big science, team science, and open science for neuroscience. *Neuron, 92,* 612–616.

LaFollette, H., & Shanks, N. (1996). *Brute Science: Dilemmas of Animal Experimentation.* New York: Routledge.

Legrenzi, P., Girotto, V., & Johnson-Laird, P. N. (1993). Focusing in reasoning and decision making. *Cognition, 49,* 37–66.

Martinson, B. C., Anderson, M. S., & De Vries, R. (2005). Scientists behaving badly. *Nature, 435,* 737–738.

McKiernan, E. C., Bourne, P. E., Brown, C. T., Buck, S., Kenall, A., Lin, J., et al. (2016). How open science helps researchers succeed. *eLife, 5,* e16800.

Merali, Z. (2010). The large human collider. *Nature, 464,* 482–484.

Musil, R. (1937). *Der Mann ohne Eigenschaften.* Hamburg: Rowohlt Verlag. Available at: https://archive.org/stream/MusilDerMannOhneEigenschaften.

NCC Staff. (2013, August 28). 10 famous quotes from Dr. Martin Luther King, Jr. *Constitution Daily.* Available at: https://constitutioncenter.org/blog/10-famous-quotes-from-dr-martin-luther-king-jr.

Northcraft, G. B., & Neale, M. A. (1986). Opportunity costs and the framing of resource allocation decisions. *Organizational Behavior and Human Decision Processes, 37,* 348–356.

Nosek, B. A., Alter, G., Banks, G. C., Borsboom, D., Bowman, S. D., Breckler, S. J., et al. (2015). Promoting an open science culture. *Science, 348,* 1422–1425.

Nosek, B. A., Spies, J. R., & Motyl, M. (2012). Scientific utopia: II. Restructuring incentives and practices to promote truth over publishability. *Perspectives on Psychological Science, 7,* 615–631.

Open Science Collaboration. (2015). Estimating the reproducibility of psychological science. *Science, 349,* 943.

Rose, N. (2014). The human brain project: Social and ethical challenges. *Neuron, 82,* 1212–1215.

Phillips, K. A., Bales, K. L., Capitanio, J. P., Conley, A., Czoty, P. W., 't Hart, B. A., et al. (2014). Why primate models matter. *American Journal of Primatology, 76,* 801–824.

Roelfsema, P. R., & Treue, S. (2014). Basic neuroscience research with nonhuman primates: A small but indispensable component of biomedical research. *Neuron, 82,* 1200–1204.

Russell, W. M. S., & Burch, R. L. (1959/1992). *The Principles of Humane Experimental Technique.* Wheathampstead: Universities Federation for Animal Welfare. Available at ALTWEB: http://altweb.jhsph.edu/pubs/books/humane_exp/foreword.

Schelling, T. C. (1971). On the ecology of micromotives. *The Public Interest, 25,* 61–98.

Singer, P. (1981/2011). *The Expanding Circle: Ethics, Evolution, and Moral Progress.* Princeton, NJ: Princeton University Press.

Singer, P. (1998). *Ethics in Action: Henry Spira and the Animal Rights Movement.* Lanham, MD: Rowman and Littlefield.

Smaldino, P. E., & McElreath, R. (2016). The natural selection of bad science. *Royal Society Open Science, 3,* 160384.

Spiller, S. A. (2011). Opportunity cost consideration. *Journal of Consumer Research, 38,* 595–610.

Than, K. (2011, May 26). Densest matter created in big-bang machine. *National Geographic News.* Available at: https://news.nationalgeographic.com/news/2011/05/110524-densest-matter-created-lhc-alice-big-bang-space-science.

The International Brain Laboratory. (2017). An international laboratory for systems and computational neuroscience. *Neuron, 96,* 1213–1218.

Von Wieser, F. (1914/1927). *Social Economics* (A. Ford Hinrichs, Trans.). New York: Adelphi.

Zhang, N., Ji, L.-J., & Li, Y. (2017). Cultural differences in opportunity cost consideration. *Frontiers in Psychology, 8,* 45.

INDEX

© The Editor(s) (if applicable) and The Author(s) 2018
J. Lauwereyns, *Rethinking the Three R's in Animal Research*,
https://doi.org/10.1007/978-3-319-89300-6